HACKING THROUGH TIME

HACKING THROUGH TIME

. . . .

FROM TINKERERS TO ENEMIES OF THE STATE (AND SOMETIMES, STATE-SPONSORED)

PEDRO B.

This book is dedicated to

My mom

No matter where you've gone,
I'm sure that you are there among the stars

CONTENTS

· · · · · · · · · · ·

INTRODUCTION

· · · · · · · · · · ·

The word "hacker" has matured through time within the internet universe. The word actually predates the internet. However, even though maturity implies change, it is not always an improvement. Three decades ago, it meant "curious" or "tinkerer," and nowadays, it mostly means "criminal," which is clearly not better.

Written for techies and non-techies alike, this book is a journey through the origin of the word "hacker." It covers how its meaning has changed since and throughout the last century, with first-person examples, thoughts, and a focus on what can and will (or not) happen next.

WELCOME

· · · · · · · · · · ·

Welcome, dear reader, and thank you for joining me in this first-person voyage through time.

> ***Please note*** *– this is not fiction (but it sure looks like it, sometimes). As such, some names have been purposely left out. But not all.*

PART I
THE 1970S

THE BEGINNING

What does "hacker" really mean to you?

The term "hack" did not originate from computers. Rather, it was derived by MIT's Tech Model Railroad Club way back in 1961, when club members hacked their high-tech train sets in order to modify their functions.[1]

Plus, the transistor was invented in 1947, so one might think it is impossible for hacking to predate that.

However, the origin of hacking is a long way from state-sponsored criminals whose job is to try and take down critical network infrastructure.

The very first hack happened in 1878 when the Bell Telephone company was started. A group of teenage boys, who were hired to run the switchboards, would disconnect or misdirect calls. From then on, the company chose to employ female operatives only.

In 1903, magician, inventor, and wireless technology enthusiast Nevil Maskelyne managed to disrupt John Ambrose Fleming's first public demonstration of Marconi's supposedly secure wireless telegraphy technology by sending insulting Morse code messages discrediting the invention.

Nevil managed to send rude messages via Morse code through the auditorium's projector and humiliate Marconi.

In a letter to The Times newspaper, Fleming asked readers for help to unmask the scoundrel responsible for such "scientific vandalism." Interestingly, Maskelyne himself replied, claiming that his intention had been to unmask Marconi and reveal the vulnerability of his invention.

1939-1945

· ·

Thanks to the code breakers at Bletchley Park, the Allies were able to read enemy intelligence reports and orders, playing a key role in the defeat of Nazi Germany.

With the capture of a German Enigma machine, the tide of World War II turned in favour of the Allies.

What do you think the Poles, Brits and Americans, who each broke Enigma ciphers at different points during the war, had as their (most likely unofficial) job title?

They weren't "Security Analysts"; they were hackers.

But even then, human error played a part – and a crucial one at that: every message ended with the phrase "Heil Hitler," which gave the Allies a baseline to infer how the cipher worked.

Therefore, one can say that vintage – albeit far from basic – hacking saved millions of lives.

I'll give you, the reader, a minute to let that statement sink in.

1957 ONWARDS

· ·

Phone hackers, aka "phone phreaks," first emerged in the US in the late 1950s. They would listen to tones emitted by phones to figure out how calls were routed.

The technique known as phreaking was discovered by a blind seven-year-old boy.

The unlikely father of phreaking, Joe Engressia, aka Joybubbles, was a blind seven-year-old boy with perfect pitch.

In 1957, Engressia heard a high-pitched tone on a phone line. He began whistling along to it at a frequency of 2600Hz – exactly that needed to communicate with phone lines and activate phone switches.

Hackers quickly exploited the discovery and used the technique to get free long-distance international phone calls.

One fateful day in 1971, a young hacker named John Draper opened a box of Captain Crunch cereal. It is hard to imagine how this led to a major event in the history of anything, but the detail is that the box came with a toy whistle. Draper didn't take long to realise he could use this toy whistle to simulate the exact tones required to make free calls. This stunt earned him the nickname Cap'n Crunch, and it also led him to create something called the blue box. The blue box was a device designed to mimic the tones used by phone companies to make free calls anywhere. They were used until the late '90s.

1963
. . . .

The first-ever reference to malicious hackers was published by MIT's student newspaper.

Following that, the term seems to have migrated from the MIT context to computer enthusiasts in general, and, in time, it became an essential part of their lexicon. The Jargon File, a glossary for computer programmers that was launched in 1975, lists eight definitions for "hacker."[2]

The first reads, *"A person who enjoys exploring the details of programmable systems and how to stretch their capabilities, as opposed to most users, who prefer to learn only the minimum necessary."* The following six are equally approving. The eighth, and last, is *"A malicious meddler who tries to discover sensitive information by poking around."*

1969

One of the biggest hacks created in the '60s, 1969 to be exact, was developed as an open set of rules in order to run devices quicker. It was designed by two employees from the Bell Lab's think-tank. The two employees were Dennis Ritchie and Ken Thompson.

The hack's name was UNIX, and it soon became one of the most popular operating systems around the globe.

1972/1975

Apple founders Steve Wozniak and Steve Jobs began building 'blue boxes' – electronic devices that allowed people to make free, illegal, long-distance phone calls. They mimicked the same 2600hz "switching" tone used by telephone operators to connect people, tricking automated systems.

They were built by Steve Wozniak and marketed by Steve Jobs circa 1972.

1970-1995

· ·

Remembered as one of the most notorious hackers in internet history, Kevin Mitnick started out with a humble interest in ham radio and computing.

From the 1970s until 1995, Mitnick penetrated some of the world's most highly-guarded networks, including those of Motorola and Nokia.

His first brush with the law came in 1981 when, as a 17-year-old, he was arrested for stealing computer manuals from Pacific Bell's switching centre in Los Angeles.

Mitnick used elaborate social engineering schemes, tricking insiders into handing over codes and passwords and then using the codes to access internal computer systems. He was driven by a desire to learn how such systems worked but became the most-wanted cyber-criminal of the time. Mitnick was jailed twice, in 1988 and 1995, and was placed in solitary confinement while in custody for fear that any access to a phone could lead to nuclear war.

He is now the owner of a security company that specialises in hacking into its clients, employing some of the techniques that Mitnick used in his previous criminal activities. While Mitnick never profited from his illegal hacking, he felt no ethical responsibility to his victims. *"So, it's kind of interesting because what other criminal activity can you ethically practice? You can't be an ethical robber. You can't be an ethical murderer. So, it's kind of ironic. But it is really rewarding to know that I can take my background and skills and knowledge and really help the community."*

These were times of legendary hacking binges – days and nights with little or no sleep – leading to products that surprised and sometimes annoyed colleagues in mainstream academic and research positions. The "pure hack" did not respect conventional methods or theory-driven, top-down programming prescriptions. To hack was to find a way – any way that worked – to make something happen, solve the problem, and invent the next thrill. There was a bravado associated with being a hacker: an identity worn as a badge of honour. The unconventional lifestyle did not seem to discourage adherents, even though it could be pretty unwholesome: a disregard for patterns of night and day, a junk-food diet, inattention to personal appearance and hygiene, the virtual absence of any life outside of hacking. Neither did hackers come off as very 'nice' people; they did little to nourish conventional interpersonal skills and were not particularly tolerant of aspiring hackers with lesser skills or insufficient dedication.

It was not only the single-minded attachment to their craft that defined these early hackers, but also their espousal of an ideology informally called the "hacker ethic."

This creed included several elements:

- A commitment to total and free access to computers and information,
- A belief in the immense powers of computers to improve people's lives and create art and beauty,
- A mistrust of centralised authority,
- A disdain for obstacles erected against free access to computing,
- An insistence that hackers be evaluated by no other criteria than technical virtuosity and accomplishment (by

hacking alone and not "bogus" criteria such as degrees, age, race, or position).

In other words, the culture of hacking incorporated political and moral values as well as technical ends.

In the early decades – the 1960s and 1970s – although hackers' antics and political ideology frequently led to skirmishes with the authorities (for example, the administrators at MIT), generally, hackers were tolerated with grudging admiration.

Even the Defence Advanced Research Projects Agency (DARPA), the funding agency in the US which is widely credited for sponsoring the invention of the internet, not only turned a blind eye to unofficial hacker activities but indirectly sponsored some of them. For example, the research it funded at MIT's artificial intelligence laboratory was reported online in 1972 in HAKMEM as a catalogue of "hacks."[3]

This report is prefaced, tongue-in-cheek, as follows:

· · · · · · · · · · ·

"Here is some little-known data which may be of interest to computer hackers.

The items and examples are so sketchy that to decipher them may require more sincerity and curiosity than a non-hacker can muster."

· · · · · · · · · · ·

Eric Raymond, prolific philosopher of the Open Source software movement, suggests that for DARPA, *"The extra overhead was a small price to pay for attracting an entire generation of bright young people into the computing field."*[4]

It is clear by now that hackers were never part of the mainstream establishment. However, their current reputation as cyberspace villains is a far cry from decades past when, first and foremost, they were seen as ardent (if quirky) programmers capable of brilliant, unorthodox feats of machine manipulation. True, their dedication bordered on fanaticism, and their living habits verged on the unsavoury.

But the shift in popular conception of hackers as deviants and criminals is worth examining, not only because it affects the hackers themselves and the extraordinary culture that has grown around them. It also reflects shifts in the development, governance and meaning of the new information technologies.

PART II

THE 1980S

During this decade, the mass marketing of the personal computer and new telecommunications technology like the acoustic coupler allowed phreaks and hackers to begin communicating via computerised bulletin board systems. Hobbyists would have a computer hooked up to a perpetually open line that anyone could dial into. The users would upload and download files, messages, programs, etc. The culture was also transmitted via the local hacking and phone phreaking communities, where there would be meet-ups and people getting to know one another in real life.

The co-evolution of the network of bulletin board systems and the ARPANET built infrastructure finally interconnected, which caused intercultural tension that continues to this day.

As the links between these hobby systems and the growing system of inter-networked computers became more complex, they evolved into what we think of as the internet today. A distinct culture was built with the interaction of computers attached to the telephone network. It was not used for its popular purpose of voice communications, but instead to transmit computer data between hobbyists and a number of entrepreneurial computer operators. These people were overlapped, intertwined, and in tension with the academic and military computer users.

By this point, computer bulletin boards had already started popping up all over the world. Hobbyists were connecting their computers to their phone lines, allowing others to dial in and leave messages, programs, and ASCII-based art. The textfiles.com archive contains 58,227 files, adding up to over a billion bytes of information.

By 1983, there was already a major motion picture – WarGames – dramatising the myth of the teenage hacker almost kick-starting the apocalypse. The Department of Justice had also already produced a report on the techniques and standards required for computer security.[5]

This document explicitly stated the position that information contained in computers was an asset, that computer security was necessary, and that it opened up companies to liability if they didn't perform due diligence in securing things like customer data and personal information, etc.

At this moment, in 1984-85, the publication of a couple of renowned magazines began to transmit the values and norms of hacker culture.

Phrack Magazine was an underground e-zine that was ostensibly published out of The Metal Shop BBS in November 1985. Craig Neidorf and Randy Tischler, known as Knight Lightning and Taran King, were the ones behind it.

Phrack was made up of "philes" or text files that covered topics ranging from the home manufacture of drugs to highly technical schematics of telephone systems and computer equipment.

The Phrack table of contents below illustrates the topics that are still discussed by the popular hacker culture to this day. From this, it's easy to see how hackers began to get linked inextricably with computer security and security in general. Often, those who were outside of universities, corporations, or the military, still wanted access to computing resources.

Some were espionage agents; some were just curious. Many hackers' first recourse was to just go ahead and access the systems without worrying overmuch about the legality. At that time, it was seen as a harmless exploration akin to trespassing.

This changed in 1986 with the passing of the Computer Fraud and Abuse Act.

The penalties became much stricter after this. Violators were generally facing decades of jail time with no opportunity for parole.

Volume One, Issue One, released on November 17, 1985. Included are:

1 *This Introduction to Phrack Inc. by Taran King*
2 *SAM Security Article by Spitfire Hacker*
3 *Boot Tracing on Apple by Cheap Shades*
4 *The Fone Phreak's Revenge by Iron Soldier*
5 *MCI International Cards by Knight Lightning*
6 *How to Pick Master Locks by Gin Fizz and Ninja NYC*
7 *How to Make an Acetylene Bomb by The Clashmaster*
8 *School/College Computer Dial-Ups by Phantom Phreaker*

The first issue of Phrack exposes an already evident tension between the technical and the antisocial. Technical documentation was juxtaposed with philes that taught the reader how to pick locks and make acetylene balloon bombs. Issue two contained an in-depth overview of MCI Communications Corporation that included data such as subscriber figures and descriptions of various services, plus philes that provided instructions for making homemade guns and blowguns. Issue four contained a phile guiding the reader through the process of making methamphetamine. In essence, Phrack seemed to live up to its mission statement; it was obviously geared toward the mischievous adolescent male.

Phrack's mission was to bring technical information to the hacker/phreaker collective with a decidedly anarchist/countercultural bent. Phrack stuck mainly to practical aspects of technology, teaching nascent hackers the tricks of the trade, but it was also a repository of self-generated hacker culture.

Phrack does not mention anything about information freedom for quite some time. This is unexpected in one of the most famous hacker cultural artefacts ever. Instead, Knight Lightning responds to the phrase in a flippant manner. It seems "information should be free" was a common truism.

· · · · · · · · · · ·

"Information shouldn't be free; you should find out things on your own."

· · · · · · · · · · ·

The slogan was so accepted that it doesn't even need to be discussed directly, though Mr Lightning seemed to have a different take on it:

· · · · · · · · · · ·

"Knowledge is the key to the future, and it is FREE. The telecommunications and security industries can no longer withhold the right to learn, the right to explore, or the right to have knowledge. The new age is here, and with the use of every LEGAL means available, the youth of today will be able to teach the youth of tomorrow."

· · · · · · · · · · ·

1984 also brought us the publication of the first issues of *2600: The Hacker Quarterly,* including a powerful example of what it meant to be a hacker.[6] Edited by the pseudonymous Emmanuel Goldstein and published out of New York, this 'zine published exploits, code, news, and pictures of payphones from around the world.

The Hacker Ethic and the generation of a culture surrounding computers necessarily began within the institutional environments ultimately provided by DARPA. A conspiracy theorist might want to argue that it was an intentionally created thing, like an intentionally seeded countercultural inoculation that trained a legion of young people in the bread-and-butter of the (then) future of espionage and spy-craft.

The ones who get punished aren't necessarily the technical wizards.

Often, those who crafted the programs and wrote the viruses passed them along via an anonymous BBS upload for others to implement and use.

The internet is a system that was designed to survive the apocalypse.

Abstract away the loss of a city to a simple node going down, but the network, and therefore, the nation, civilisation, and humanity, survives. The laws that we have in place now were created during this time when the US was involved in a war of espionage and secrecy versus foreign powers. The people dialling into military or industrial systems seemed to be domestic enemies of the state. Some claimed the ability to subvert everything that held the social order together. And even today, governments and corporations are routinely penetrated by unwanted intruders that the public believes are called hackers.

1986 – THE HACKER MANIFESTO

While reading the Phrack e-zine that could be found in all BBSs I accessed in 1988 and beyond, I came across an article in Volume One, Issue 7 that would later become widespread and, I daresay, famous: *"The Conscience of a Hacker,"* also known as The Hacker Manifesto.

This essay was written after the author (Loyd Blankenship, known as The Mentor) was arrested, and it was written on January 8, 1986.

It was a very emotional speech, and I quote:

· · · · · · · · · · · ·

"This is our world now . . . the world of the electron and the switch, the beauty of the baud. We make use of a service already existing without paying for what could be dirt-cheap if it wasn't run by profiteering gluttons, and you call us criminals.

We explore . . . and you call us criminals.

We seek after knowledge . . . and you call us criminals. We exist without skin color, without nationality, without religious bias . . . and you call us criminals.

You build atomic bombs, you wage wars, you murder, cheat, and lie to us and try to make us believe it's for our own good, yet we're the criminals.

Yes, I am a criminal. My crime is that of curiosity. My crime is that of judging people by what they say and think, not what they look like.

My crime is that of outsmarting you, something that you will never forgive me for.

I am a hacker, and this is my manifesto. You may stop this individual, but you can't stop us all . . . after all, we're all alike."

· · · · · · · · · · · ·

When asked about his arrest and motivation for writing the article, The Mentor said:

• • • • • • • • • • •

"I was just in a computer I shouldn't have been. And [had] a great deal of empathy for my friends around the nation that were also in the same situation. This was post-WarGames, the movie, so pretty much the only public perception of hackers at that time was 'hey, we're going to start a nuclear war, or play tic-tac-toe, one of the two,' and so I decided I would try to write what I really felt was the essence of what we were doing and why we were doing it."

• • • • • • • • • • •

The Mentor conjured an image of a highly intelligent youth that has been left behind by an education system that caters to the lowest common denominator. He proclaimed that his crime was "curiosity," a theme that is common to most discussions of what it means to be a hacker. The Hacker Manifesto prescribed a way of being in the digital age. It was an important indicator of the norms of a collective hacker identity, putting forth such ideals as intelligence and a hunger for knowledge.

His words echoed what I felt inside. No, I had never done anything illegal, but I had a curious mind. I related very deeply with the *"We exist without skin color, without nationality, without religious bias"* because, at the time, the internet was as inclusive as it ever was.

In that regard, the evolution of the internet has always been a step back. There were no high-quality selfies to serve as account avatars. And it was enough. As The Mentor said, people were worth their actions, thoughts and words, not what they looked like.

But the world's perception of hackers was already being driven by the media, albeit accidentally. But it was still wrong.

Even though, as we will read in the next chapters, that did eventually happen.

I can't help but think that maybe the WarGames-inspired view on hackers gave birth, as collateral damage, to what was initially just a sci-fi movie character.

Phrack was pronounced dead around 2006. Despite previous occasions in which the pronouncement of death may have been premature, it seemed that this time the pronouncement had been made not only by the editors of Phrack but also by the hacking community. One prominent member (Dark Sorcerer) asked:

• • • • • • • • • • •

"Is Phrack more or less popular than it was five years ago? Ten years ago? I don't know. It does seem as though Phrack has followed a classic organic cycle: a naive, exuberant youth paving the way for a stodgier, more establishment-minded adulthood. That's not to say that it's irrelevant, but rather that it was doing what it should have. Evidently now – whether due to exhaustion, boredom, or just plain realising it's time to move on – someone has decided to give it a rest. Twenty years was definitely a good run – so RIP, Phrack."

• • • • • • • • • • •

1986
• •

The first time I touched a keyboard (more accurately, touchpad keys with a keyboard membrane) was in 1986.

A remarkable piece of engineering – the ZX Spectrum 1k.

This lovely piece of kit was expensive, but as my uncle worked as a mechanical engineer at Timex, he picked out broken hardware from the trash, took it home, fixed it, and gifted me one.

After typing something like 20 lines of BASIC, it was "out of memory," so the 1k was quickly upgraded to a 16k and then a 48k. I do miss the rubber keys.

Well, not just one computer. Because I kept it on for many hours, I literally melted a couple. Then my uncle created – by hand – the first Spectrum 48k I've ever seen with a heat sink.

And I kept it on from morning until late in the night.

I'd spend a big chunk of my allowance buying Crash magazine, and then spend hours frantically typing in pages upon pages of assembly code. Then more hours trying to find why the code didn't run. *"Ah, a comma was missing on line 791."*

I was already ahead of what my uncle could get from the trash and fix.

Cue the natural (and expensive) upgrades to a 128k+1, +2 and +3 – my mind was in "absorption mode."

The door to a new universe had been opened, and there was no turning back.

BBSS

I have had a modem since 1988. A lovely 300 bauds at first, then 2400 bauds – yes, 2400 characters per second (so please stop complaining about your "slow 10 Mbits/s internet"), and there were not many places to connect to.

The '80s-'90s and the BBSs (Bulletin Board Systems) changed that. Many lists of BBSs were being shared, and as soon as you had a phone number to connect to a public BBS, it would be quite common to find a list with more numbers available on that BBS, and so on and so forth.

Dialling into a BBS felt like sci-fi-like teleportation. It was the intimacy of direct, computer-to-computer connection that did it. To call a BBS was to visit the private residence of a fellow computer fan electronically. BBS hosts had converted a PC – often their only PC – into a digital playground for strangers' amusement.

Maybe it was because the system operators (aka *sysops*) that ran each BBS were always watching. Everything users did, scrolled by on their admin screen, and they soaked in the joy of someone else using their computer. It was also a gentle, pleasant form of surveillance.

The sysops might initiate a one-on-one chat at any time. Long before texting and Slacking and Facebook messaging became the norm for interchange, BBS chats felt like being with someone in person. Sometimes strong personal relationships were built. I remember waking up at 6:30 am, right before my parents went to work, just to chat with a sysop that was almost always awake (and chatty) at that time.

That personal connection was sorely missing on big-name online subscription services of the time – Prodigy, CompuServe and AOL. Even today, the internet is so overwhelmingly intertwined that it doesn't have the same intimate feel. Once the web arrived in the mid-1990s, it seemed inevitable that the BBS would die off.

However, even today, a small community of people still run and call BBSs. Many seek the digital intimacy they lost years ago. According to the Telnet BBS Guide[7], 373 BBSs still operate, mostly in the United States. Many are set up to be accessible via internet-connected tools like Telnet, a text-based remote-login protocol originally designed for mainframes.

Visiting an old BBS still running today feels like strolling through a community frozen in time. The message threads are incomplete, with discussions left hanging. There are bulletins that post stern-sounding rules from the 1980s like, *"USERS WITH FAKE NAMES WILL BE BANNED FOREVER"* or *"Attempts to tamper, damage, or defraud this system are against Oregon and Federal laws and will be reported immediately to authorities."*

That sort of thing scared people back in the '80s.

The idea of having only a handful of people (or even just one person) able to connect to your server at a time is somewhat preposterous nowadays (at least in most environments). Yet that's the way it was back then. And when there were no more modems available? You'd get a busy signal – a noise that kids of today have probably never even heard.

These BBS servers handled email that was not unlike the email of today. The primary difference is that instant delivery of email occurred only on the local BBS system.

If you sent an email to a user on a different BBS, that user wouldn't be able to read that email until the two BBS systems performed a regular (often once per night) link-up to trade emails with each other.

Each such connection between two BBSs was considered a single "hop." Often, in order to get an email delivered to a physical location that was very far away, multiple such hops were required. The email would be delivered to one BBS on the first night and then to the next BBS in the line each subsequent night.

With this system (which was incredibly popular in the 1980s and into part of the 1990s), email was not an instantaneous thing. It was not unheard of for email delivery to actually take longer than postal mail. This is not a joke.

Still, millions of people used such services for email. At its peak, the most popular such network of BBSs (known as FidoNet)[8] consisted of over 39,000 dial-up bulletin board systems across the world.

But I soon found out that not every BBS was public. And I just had to know more. So, I explored. I mean, why wouldn't you want to know more? If something is locked away, even if you don't know what it is, you're curious, right?

Oh, here's the "curious" word again. As the reader probably already knows by now, curious really means trouble. At the time, I kind of didn't.

Some BBBs had a NUP – a New User Password. They weren't even complex passwords but simply served as a way to separate "those who knew" from "those who didn't."

Most of the BBSs were simply conversation forums with very archaic – albeit fun – games in what were called "Doors." People woke up (or, at least, I did) to play my daily round of "Trade Wars," which was first released in 1991 and lasted way beyond 2010, with a 2002 version released in the meantime.

But there was something else kept behind those doors in some BBSs: Warez.

I do not wish to dwell on the multiple definitions of the word, so let's just go with *"Software distributed illegally, which has had its protection codes deactivated."*

A shorter version: *"Pirated media."*

At the time, it was about so much more than the files or the content. It was about an underground pre-internet subculture.

BBSs (and subsequently, FTP servers) that were part of the "scene" were very sought-after and a small part of what was still, by itself, a small online presence.

The "scene," as it was known, was highly illegal in almost every aspect of its operations.

But it was well organised. It had leaders, managers, rules (which are still being worked on, according to scenerules. org),[9] councils, traders, couriers, rippers, crackers . . . and, over time, at least a dozen definitions of responsibilities and roles.

Some – if not most – of the internet today is not as well organised as the scene was.

This was not about releasing a file and making sure you had the biggest number of "likes" or "thumbs up." It was about sharing something with a very (supposedly) tight-knit group of "elite members" and making sure no one from the general public could touch it. Surprisingly, the "everything should be free to everyone" only came much later.

Think of a well-oiled machine, encompassing operations and infrastructures with its own norms and rules of participation, forms of sociality, and artistic forms.

There were no emojis or memes, as everything started out as text. But I have seen ASCII (and ANSI, its coloured evolution) art that was more than 10,000 lines long. And it was beautiful.

I have met many people who dislike terminal windows because *"There's no point and click with the mouse; you have to type everything."*

My obvious reply is almost always only echoed in my mind and not in words: *"Yes, that's how it all started. You are spoiled."*

1988 – THE MORRIS WORM

As a graduate student at Cornell University in 1988, Robert Morris created what would be known as the first worm on the internet, and he did it solely to give himself an idea of the size of the web.

The worm was released from a computer at MIT in 1988 in hopes of suggesting that the creator was a student there. It started as a potentially harmless exercise but quickly became a vicious denial of service attack. A bug in the worm's spreading mechanism led to computers being infected and reinfected at a rate much faster than Morris anticipated. By the time he realised the issue and attempted to rectify it by telling programmers how to kill the worm, it was too late. Once discovered as the author of the worm, Morris became the first person to be convicted by jury trial of violating the Computer Fraud and Abuse Act.

By this time, it no longer mattered if you had a curious mind or not – even mistakes like Morris' could, and would, be punishable.

However, many things were still not a crime, and I distinctly remember sending my first email. No, it was not just a couple of mouse clicks, as emails were actually a complex task. Even today, if you ask someone to send an email only by using SMTP commands using Telnet on a terminal, you will see that it has become what I can only describe as a lost art.

Nowadays, the question is, *"Why would you do something so geekingly complex to send an email?"* The simple answer is, *"Back then, you had to."*

1989 - RANSOMWARE

The first documented and purported example of ransomware was the 1989 AIDS Trojan, also known as PS Cyborg.[10]

Harvard-trained evolutionary biologist Joseph L. Popp sent 20,000 infected diskettes labelled "AIDS Information – Introductory Diskettes" to attendees of the World Health Organisation's international AIDS conference.

But after 90 reboots, the Trojan hid directories and encrypted the names of the files on the customer's computer. To regain access, the user would have to send $189 and another $378 for a software lease to PC Cyborg Corp. at a post office box in Panama.

Dr Popp was eventually caught but never tried for his scheme as he was declared unfit to stand trial.

PART III
THE 1990S

The '90s were a great decade, as far as the expansion of the internet is concerned.

Kevin Mitnick was arrested in February 1995 and held without bail. That's why I was wearing a "Free Kevin" t-shirt at the time.

Kevin represented the "true" definition of the word hacker: a curious mind, who would go to great lengths to expand what could be made possible with technology, and I shared that mindset and vision.

Some people tried to ridicule Mitnick because he was caught, saying that a real hacker should be able to cover his or her tracks well enough to evade detection and capture.

However, most of those people had delusions of grandeur, thinking they were more capable than Mitnick.

The Kevin Mitnick case is an example of how hackers have been ineffective in resonating with the general population. Part of the reason for this ineffectiveness can be summed up in three words: "Free Kevin Mitnick." And yes, I know I just said I had the t-shirt.

Although this is a good slogan, it leaves some major premises unspoken. Hackers do not contend that Mitnick is innocent, only that he is being treated unfairly and that the punishment is unjust. But why should the general public wish to release a repeat offender?

They cannot relate to a repeat-offender hacker who has been imprisoned. Perhaps if the public had a clear idea of what Mitnick had truly done, and if the reported costs of his crimes were actual rather than projections by the corporations that Mitnick had digitally entered, they might take a more sympathetic view of the Mitnick case.

In order for the general public to understand exactly what Mitnick had done, hackers would have to do more than hack webpages and post slogans; hackers would have to educate the masses. But hackers believe that the masses are stupid and unable to be educated on technical matters. Like propagandists, hackers do not have to change their opinion of the masses to change the public's opinion of hackers. Hackers could view it as an exercise in social engineering: attempting to create resonance where there should be none. So long as hackers continue to engage in protest actions that do not resonate with the general public, they will remain relegated to the fringes and thus more vulnerable to persecution and legislation that defines them in problematic ways.

The "Free Kevin Mitnick" movement illuminates current rhetorical theory, specifically the ego function of protest rhetoric, because it does not fall neatly into the rubric of self- or other-directed social movements. Although Kevin Mitnick was a hacker, "free Kevin Mitnick" was not all about hackers. Mitnick acted figuratively as a representative of the hacker movement, serving as the most visible example.

FREE?

Even though the first BBS (Bulletin Board System) dates from August 1973, modems were slow, expensive, and not available to everyone. Plus, they had added phone call costs.

At the time, information was not for all. The internet wasn't always on – one had to connect, and the process was very prone to failure. I remember memorising the Hayes set of modem commands to fine-tune it.

Even though the *"Information Wants to Be Free"* adage was coined in the '80s and was slowly but surely finding its way to every corner, not all information was free. Not because it had a cost, but because it was really hard to find.

Since its inception, and after talking to several "digital rights activists," I doubt their reasons for why all information "wanted" to be free.

What they wanted (and still do) is "free information." Or so they say.

A slightly different version is *"All information should be free, and any proprietary control of it is bad,"* which, according to Wikipedia, is part of the "Hacker ethics." However, this sounds like a utopia taken from an anarchistic guidebook.

It's true that many of the first hackers were hippies simply because of the intersection with the 1960s, but it's time to let go of (i.e., no longer the time or place for) those mannerisms and protestant ideals.

Some people have adapted the hacker culture tenets to give them the self-proclaimed right to access any information that they seek without barrier and apparently without regard for others' ability to have that same access.

This idea has since expanded to suggest not only the right but also the moral duty to actively liberate information so that it can be free for anyone to access.

I believe we all need active users, such as hackers, who are exploring the limits and edges of the system to balance against the powerful interests that claim ownership and control of the networks. Hackers have historically driven innovation in unexpected ways.

Hackers are a popular counter-cultural movement that has had an impact on the way almost everyone interacts with the world around them. It follows that the culture has successfully evolved in an increasingly hostile world.

Hackers are credited (or blamed) as the first people that

leveraged the potential for interactive computing. Because of the hackers' desire to wrangle directly with the computer, the nature of work has changed for many people all over the world. This subculture has since bloomed into a new cultural niche.

The problem with this bloom is that the core values held by this culture have clashed with normal cultural values time and time again. Information has never been free in a cultural environment of war, secret weapons, military intelligence, trade secrets and intellectual property. When most people with political clout see ideas as capital investments, it's difficult to legitimise the notion that these ideas should be freely distributed.

If industrialists had their way, we'd be ignorantly buying at the monopolistic company store, without even asking if there was something more. Sadly, this is how the vast majority of people tend to interact with computing technology as a whole. Extensive computer industrial empires have been created based on the general ignorance of the public.

If you browse Twitter, Facebook, Instagram, or other mainstream giant websites, how much "Information" can one really see? There's a lot of data, but "Information" should be synonymous with science, knowledge, and culture – and it really isn't.

What one thinks is "Information" might just be a stream of mostly mindless and unstructured data to another person with different interests. That's why someone like Elon Musk has so many followers, right?

Some are avid fans, and some think he is polluting the platform he intends (or intended) to buy with the excuse (or justification) of free speech. I am pretty sure free speech is not a synonym for "trash," but Elon's mind works in mysterious ways, to say the least.

This endless stream of trash is created by people who think they are interesting and knowledgeable enough to share what's on their minds, and that you (and everyone else!) will enjoy it. But let's face it; not everyone has something interesting to say.

The internet was born as a privilege, not a right. And as much as I have to respect my fellow humans, more than 90% of what I read on the internet today is just trash. Some of it I can avoid, but most of it is spoon-fed, unregulated, and unstructured, and I wish there was a way to *unsee* it.

Around 1991, as part of my mostly passive observation of the BBS underground, I had to choose a nickname for myself – a handle. For the last 30 years or so, a handle has meant nothing. Everyone has multiple nicknames, chosen at random. But back then, it wasn't like this. Your real name was supposed to be protected and not divulged, so the handle had a much greater meaning – it was your persona, and it would be what defined you within the internet realm.

Some BBSs also allowed for a tagline to be added to the handle, which also helped define someone's online persona. I will never forget AssKicker's *"Born to kick ass and chew C4"* tagline. And in case you are wondering, no one referred to themselves in the third person – that stupid idea was born on LinkedIn.

I got a nickname in 1991 that I still use. Well, not for every website where I have a login because, after 30 years and billions of people joining the fray, it's close to impossible to have a unique nickname.

Nowadays, "Nickname taken, choose another" is common, but it still irks me a bit to see that "someone else" created an account using a nickname that has been "mine" for decades.

TO BE 1337 – OR NOT

. . . .

The early '90s is sort of when *leetspeak* came in – at least for me, but it had existed since the '80s.

Filters were often used by BBS administrators in order to ban the use of certain words. For example, if users in chat rooms wrote about "hacking" or "cracking," the filters would block the content. However, the "elite" of internet users – comprising programmers and coders – were interested in discussing these exact topics.

And so, "elite" BBS users invented *leetspeak* as a sort of cipher. On public boards and chats, *leetspeak* was used to talk about nefarious topics that went against the rules. Yes, dear reader, some BBSs censored content – around 50 years before Facebook ever did.

To circumvent the filters, users developed *leetspeak* by replacing letters with similar-looking numbers and characters. Filters were easily able to detect and block banned words like "hacker" or "ass," but they had difficulty identifying "H4x0r" or "@$$."

Today, what is a very crappy way to make your password become something that you think is more secure (spoiler: not really), at the time, was just a way to evade filters and look (sort of) cool while doing it.

It wasn't a "geek thing"; it was the norm. Doing it now, well, that's a whole different thing. 1337 skills are not a good thing anymore, but it is still an excellent example of how human creativity, armed with only an 80x25 characters terminal as a "weapon" (yes!), improved its ways of communicating. With no incessant stream of memes and graphics, it was tough.

Speaking of memes, the first ASCII emoticons are generally credited to computer scientist Scott Fahlman. He proposed what came to be known as "smileys" – ":-)" and ":-(" – in a message on the bulletin board system (BBS) of Carnegie Mellon University in 1982.

For the record, I still prefer smileys to memes, most of the time. :)

I missed the Galactic Hacker Party Conference in 1989 because of two simple reasons: I had no idea it happened, and I was too young.

As I said, this was all before the internet – email and messaging were extremely limited. And you needed to have your eyes and ears pointed to the right place, so to speak, to know about these things.

1992
· ·

On a rainy morning in December, I realised some people were missing from my morning uni class. It wasn't random – I knew most were all high-profile BBS users; some were even part of the *International Network of Crackers* (INC) group.

I later found out that *"The police took everything, only left a joystick."*

And there we had it – the first raid/bust happening very close to me.

At the time, I was only a watcher – I had access to some BBSs, and I could "see" some cracked software being traded, but I wasn't involved.

Of course, I went to local gatherings and lunches, and I knew a lot of people – mostly by their nicknames. But seeing this happen to individuals I considered my friends – uni colleagues – was scary.

I kept a low profile, yet I couldn't simply stop being curious. But I trod carefully.

In 1993, I missed the *"Hacking at the End of the Universe"* conference because I only knew about it after it happened. However, I became aware that "something" would happen every four years, so I kept waiting for 1997.

MORE HACKERS

But before jumping to 1997, we have to make a stop in 1995, literally, for Hackers' sake.

The movie *Hackers* was released. Why is that important?

On the back of William Gibson's Neuromancer from 1984, which marked the beginning of the term "cyberpunk" and outlined the content and ethos of the cyberpunk literary movement that followed, for me (and many others), the movie was so much more than that.

Let's see:

- The cast was phenomenal. I'm talking about the whole cast, not just Angelina Jolie. Even though, at the time, that was important to a young adult like me.
- The soundtrack was outstanding. Music can – and often does – mark the generation listening to it, so it did leave a mark. And it wasn't just a fad – as of today, I still think most of the tracks are on my all-time top 50.
- The characters' nicknames are forever etched in the internet hall of fame. Ok, maybe I'm overdoing it, but

"Zero Cool"/"Crash Override," "Acid Burn," and "Cereal Killer" are still common in many (too many) websites, where people create accounts thinking no one heard these nicknames before. It might be a tribute, but it's really far from being original.

Apologies to "The Phantom Phreak," "Lord Nikon," and "The Plague" – I left you out on purpose. Can't please everyone, I guess.

- The plot was good enough. "Nice" hackers vs. "mean" hacker, fun, and the "nice" hackers win.
- Critics gave the movie a really harsh review. Why?

    Because hackers were entertaining? Because the critics had no idea about what exactly a hacker subculture was? Or maybe they were afraid that hacking had become "cool."
- The movie quotes *"Hack the Planet!"* and *"Mess with the best, die like the rest!"* are still in people's minds after almost three decades.

While the plot was pure Hollywood, the film's attention to detail about hackers and hacker culture helped it gain the attention of some hackers in the underground. Of all the films about hackers, *Hackers* makes the most concerted effort to portray the hacker "scene" in some detail. It even went so far as to get permission from Emmanuel Goldstein, the publisher of the hacker quarterly *2600*, to use his name for one of the characters in the film.

These hackers are not isolated loners or misunderstood teens; they are cutting-edge techno-fetishists who live in a culture of "eliteness" defined by one's abilities to hack, phreak, and otherwise engage technological aspects of the world (including pirate TV and video games).

Hackers themselves have occasionally documented their own culture in an effort to resist media interpretations of their activities.

Hacker style is manifested in the wardrobe of the hackers. While several characters dress in typical teenage garb, the two lead hackers (played by Johnny Lee Miller and Angelina Jolie) prefer a high-tech vinyl and leather techno-fetish look. Miller's character (Dade Murphy, aka Zero Cool and Crash Override) and Jolie's character (Kate Libby, aka Acid Burn) serve as representatives of the hacker-elite sense of style.

Their look was urban – very slick and ultracool. Still is.

1995

Also in 1995, Russian computer programmer Vladimir Levin managed to steal $10 million – but not by going online. He hacked into the Citibank telephone system and stole customers' account credentials (passwords and account numbers) when they told them aloud to service reps.

Levin then used those credentials to electronically transfer millions to various accounts around the globe. He was eventually caught and sentenced to three years in prison. All but $400,000 was recovered.

This was one of the first high-profile and public electronic thefts from a financial institution.

1997 – HELLO, NETHERLANDS

A friend of mine shared an email with me. The subject was:

. . . .

"Announcing HIP97: A hacker convention and festival in the Netherlands on the 8th, 9th and 10th of August 1997."

. . . .

The email read:

. . . .

"HIP is an acronym for 'Hacking In Progress.' It will take place on Friday 8th, Saturday 9th and Sunday 10th of August 1997 at campsite Kotterbos, Aakweg, Almere in The Netherlands.

Hundreds of hackers, phone phreaks, programmers, computer haters, data travellers, electro-wizards, networkers, hardwarefreaks, techno-anarchists, communications junkies, cyber- and cypherpunks, system managers, stupid users, paranoid androids, Unix gurus, whizz kids and warez dudes spent three days building their own network between their tents in the middle of nowhere.

HIP97 will happen on the same days as Beyond HOPE, a hacker convention in New York, organized by the people of 2600 Magazine. There will be audio and video links between both events, and we're working on cool gadgets to further enhance your sense of 'grassroots telepresence.'"

. . . .

So, like Twitch, but in 1997.

• • • • • • • • • • •

"Ring the alarms. Sound the bells. Or is it the other way around? Whatever, I don't care. Let's go!"

• • • • • • • • • • •

Well, it wasn't as easy as that. The only PC I had at the time was a full tower, and my monitor was a CRT, so we had to pay a lot of money to get the 70kg/154 pounds flight case on a plane.

And then we had to carry it to the campsite. No, of course we didn't have a car. Too young and broke for that. Our money was for beer.

But I clearly remember some folks helping us with the case when we got off the bus.

'What a friendly bunch of hackers,' I thought.

As part of the check-in, we had to fill in a form with our handle so we could have a *<handle>@hip97.nl* email. Cool.

My handle had nine chars, and the limit was eight, so I had to phoneticise it. Yeah, this was 1997 – eight chars was not only the norm but also the limit to what technology could achieve.

A personal note on eight chars being the very insecure limit back then. If I ever find a way to go back in time, I'd like to land in 1997 and try to hack every single password known to man(and woman)kind at the time. That would be fun.

Now weaponised with a *@hip97.nl* email we could show off, even though we only had less than 20 email contacts, we moved on. We picked up a peg (yes, a peg!) with an IP written on it that we would use for the next three days.

There were wooden tables inside the main gigantic, circus-sized tents. Tower and CRT plugged in; IP configured quickly. Game on.

Next – obviously, an internet speed test. "So fast!"

I later found out that the uplink was 2x6Mbits. Groundbreaking. Hey, reader, it was 1997, so please stop laughing. It was very fast.

My CRT broke on the second day, and I was very worried I'd miss most of the conference. But a friendly local gentleman gave me a ride to a local hardware store, and I bought a cheap CRT which would last for years. I offered him some money for the ride back and forth, but he only wanted a beer. Or three. I wasn't counting.

Throughout the event, this spirit continued – people supporting each other. I remember going around with a bunch of fibre optic cabling, helping people connect to the hubs and switches that were scattered all over.

Everyone was connected, and it just felt "good" to be a part of it.

In total, there were 1500-2000 people, many (500ish) tents outside, and around 1000 PCs connected.

Some people had brought their shiny, new Windows OS to the conference.

Maybe they shouldn't have. There were a lot of Blue Screens of Death going on.

I think it was only then that I truly realised who was around me – more than fifteen hundred hackers.

'But they're nice folks,' I thought.

And then the organisers told us that police would be going around, and the only way to know who they were was to look for the orange badges (mine was blue).

So . . . the police were looking for wrongdoers. Does that mean that there were people doing illegal things?

In a nutshell, yes. The BSODs that I mentioned were happening as pranks. Because everyone had a peg with their IP, it was easy to find the IP of the person sitting next to you.

Even though I had a Linux (specially installed to be able to endure a tough crowd), I still hid my peg.

I remember chatting with the guy who cracked PGP 5.0; he was there. (Before you ask – it's Windows swapfile stuff, and the passphrase was stored in memory in plaintext.)

The gentleman sitting on my left was shy, and he had some trouble with his machine. It was a beautiful SPARC with SUN's Solaris 9, 500MHZ UltraSparc IIe processor and 256MB RAM. At the time (you guessed it), that hardware was fast.

We gave him a hand, and he paid us with the only currency he had: a huge bag of weed. We didn't ask for any compensation, but he really wanted to share.

There was a mock epitaph to Bill Gates near the centre of the camp. *"Where do you want to go today?"* asked the engraving. The graveside was littered with offerings – few of them complimentary.

I quickly need to mention Jolt Cola. I had never tried it before, and at 160mg caffeine per can, 190 calories and 50g of sugar, along with another 50g of "added sugar" (that's two Mountain Dews, three Dr Peppers, or three Cokes), it was everyone's beverage of choice. Beer was a close second, obviously.

Jolt Cola is, unsurprisingly, not around anymore. But in the three days of the convention, I must have slept eight hours in total. Tops. That thing really worked.

1999
. . . .

In 1999, *the scene* was everywhere. Even though there was still an attempt to keep it limited to only people "in the know," it was hard to escape the fact that what was once a secret was starting to become the norm. I wasn't too happy about the fact that people who had recently bought a modem could easily become part of the scene.

What nowadays is known as catfishing, at the time, was just the way for a scene newcomer (aka noob) to be involved in what once was an elite (and elitist) group of people.

However, you either tag along or become obsolete.

There are exceptions, of course. It's now 2022, and I refuse to create a TikTok account or even look at it. The gross monetisation of people's stupid antics makes me cringe. It's bad enough there's a TV channel somewhere that is always showing a Kardashians episode.

But what do I mean by "tag along?"

I have always been a lover of good music. All genres, as long as it's good.

And, being a collector, I owned some pretty rare things, music-wise.

So, around 1999, because I knew someone who knew someone who knew a lot of people, I decided to share a set of (rare) music files I had in my possession.

Amazingly enough, because "the scene" was becoming so accessible to "non-sceners," my release hit the – at the time – public sites that everyone had access to.

Lo and behold, the band saw it, and they made an (also public) comment to *"Thank the person or group who had helped in sharing the news about their first album."*

I was aghast. They actually thanked us (well, me) for sharing their music with the world. Obviously, there was no Spotify at the time. No YouTube. Bands had no real way to make themselves known apart from traditional marketing, which was expensive.

As I have always been a fan of *"I'm ok with sampling something, and if it is good, I'll buy it,"* I went along with it – for a bit.

Everything else on the internet was expanding at a frantic pace. Both good and bad.

For the first time, I had the feeling that those anonymous internet handles could be someone I knew, or at least someone that knew someone I knew.

It was as if the internet was no longer this ethereal realm of cables, bits and bytes – it had evolved into the real world, and that was happening all around me.

Although still, to this day, there are many that only dwell in online activities by being alone, many groups of like-minded individuals were starting to gain structure and momentum.

There were too many to name them all, but my one mention is the first (h)activist group I encountered and knew a couple of members from. *KaotiK*, founded around 1997, were a very active group in the human rights East Timor campaign. They hacked and defaced numerous Indonesian websites, created the first e-zine about hacking & security, and were dedicated to the Portuguese people.[11]

1999 – THE MELISSA VIRUS

Right before the year 2000, David L. Smith from New Jersey masked a virus as a simple Microsoft Word attachment to an email. Once the unsuspecting recipient downloaded the file, the virus replicated itself and sent out copies to the first 50 names in the victim's contact list.

Some estimates claim that roughly 20% of the world's

computers were infected. However, no sensitive information was stolen, though many businesses were disrupted.

Eventually, some businesses had to restrict internet access or shut down their email networks, including Microsoft, Intel, Lockheed Martin, and Lucent.

In the end, David L. Smith served 20 months behind bars and caused an estimated $80 million in damages in lost productivity.

1999 – NASA

Also in 1999, NASA had to shut down computers that supported the international space station for 21 days after its network was accessed by a 15-year-old hacker, James Jonathon, who went under the name of 'cOmrade.'

After first being introduced to computers at the age of six, where he would simply play games, his attention soon turned to learning about programming languages and operating systems. After his parents became worried about his obsession with computers, they banned him from using one. He insisted that his time on the computer wasn't affecting his high grades at school. No wonder he was so sure – it was later revealed that he had hacked into his school's network to change his scores.

After installing malware on a server in Alabama, he was able to compromise 13 computers on NASA's network. He acted as an employee, downloading source code which controlled the life support systems on the ISS, the worth of which was estimated to be in the millions. After discovering this, NASA disconnected the server and infected machines for three weeks to find the cause of the intrusion, costing them roughly 40 thousand dollars.

Cue the 2000s. Things are about to get weird(er).

PART IV

THE 2000S (NOW)

As soon as the 21st century dawned on us, it started to become clear that the internet-wide mindset had changed, and hackers were playing a new role.

The Y2K bug was mostly a no-show because techies prepared for it. I remember spending New Year's Eve at work, and then nothing happened, apart from everyone getting drunk. I mean, if you're working to fight a software bug, you might as well have some fun, right?

I was still involved with the MP3 scene and, later on, the SVCD scene. Yes, there was such a thing. Actually, dear reader, if you look at the *Scenerules* website,[12] you'll see that there are rules for almost every single type of software you can think of.

And this is where I started realising that what had caught my interest in the '80s was no longer true. Or that it had evolved into something nefarious.

FREE SPEECH
. . . .

Free speech started to become a thing, uttered left and right, serving as an excuse to say anything.

Much has been said about it, and it still hasn't been addressed properly. There's an enormous grey area between defamatory, insulting speech and what is considered to be everyone's right to their opinions.

People are judged by whom they follow and "like" on Twitter. And the gap is real: let's imagine that, on Twitter, I follow Manning, Snowden and Assange, while my next-door neighbour follows Trump and his devotees.

Just by knowing whom each other follows and likes, the basis for a street brawl is already in place. And, if you ask both parties, perfectly justified. Because the other person "is wrong."

At the end of the day, it's a good thing that most of today's youth are too lazy to engage in a face-to-face word confrontation.

Why do it if there are 3633 emojis available?[13]

PC WORLD MAGAZINE, JULY 2000

A couple of months after the Y2K bug no-show, I started to become aware that companies weren't really doing as much as they should in order to protect themselves from cybercrime. Yes, 22 years ago, this was already a problem.

I got in contact with a reporter from an at the time very well-known magazine (PC World). There was interest from their side to know more about the hacker community, hacker interests, and hacker mentality.

Game on.

I laid out some ground rules: We would have dinner in a location chosen by me, no names mentioned (only nicknames, if they accepted to be mentioned in the article), all pictures blurred, and I would review the final wording before going to print.

And that's exactly what happened. I still remember (with a grin on my face) the fact that no one cancelled at the last minute, which was a good change from the norm. Pictures were taken (of course, I wore my HIP97 t-shirt), just slightly blurred to be recognised by those who attended and no one else.

The topics discussed were varied. Groups (present at the gathering, mentioned in the previous chapter) briefly discussed how they were disrupting Indonesian East Timor operations and websites; how a company that paid users for website clicks was vulnerable to a script impersonating humans; how it was so easy to access a company's mail server and check if they were vulnerable to Open Relay; and last but not least, how the undercover cyber law enforcement should get more training, and better hardware.

The reporter made a positive remark about how inclusive we were in terms of gender, age, and race. I believe she was both surprised and happy to see how diverse we were. Don't forget; this was the year 2000.

All in all, nice conversations were had, and I felt that all topics covered would be useful to the general public.

Unfortunately, not quite.

The first indicator of possible trouble was two-fold. The article made the cover of the magazine in July 2000, and the title was *"This man can enter your PC."* And there I was. It was a blurred photo, but me.

I thought that title was just a bit too . . . flashy. And, ultimately, not true. I couldn't just hack any PC in the world, right?[14]

The aftermath after the magazine came out was devastating.

Many, many – too many – people that knew my nickname *"From their internet neighbourhood,"* which at the time was mainly IRC, directed their anger at me. From "media wh**e," to "snitch" and "sell-out," I heard it all. And then some.

Most of them were clearly envious of me being in a cover article. Others were mad because I had a positive view of law enforcement. Others did not accept the fact I exposed how easy it was to use an open-relay mail server to send what were, at the time, untraceable emails.

Others thought that the magazine paid for dinner and "bought" our statements. No, they didn't. In total, we were around 12 young adults, definitely not rich or anything close to it, and we all paid for our food and drinks. Which, to some, me included, was stretching the budget.

This was my first experience with internet bullying, and I was the target.

So much for my good intentions. From that moment onwards, I decided that I would never be in the spotlight again, even if I was trying to do the right thing.

Even today, I mostly think that the "general internet population" is mostly unworthy of good things. The default is still to either rage at anything and everything (e.g., Twitter, where enragement means engagement), or make up positive things by manipulating crowds (e.g., influencers of Instagram).

And I don't want to be part of it.

To be quite frank, seeing as to how some hackers hit the news in the last decade by being snitches like *Sabu*[15] or righteous yet loud like *th3j35t3r*,[16] being in the audience now gives me the perspective of how things can be perceived in a totally different way. I now think that maybe I was too ahead of the times for that kind of interaction with the public.

2001 – HAL –
THE NETHERLANDS (AGAIN)

Hackers At Large 2001 was held in Twente: a fantastic venue.

I attended and noticed things had evolved a lot since HIP – 15km of Category 5 cable for the ethernet backbones and a 1Gbit internet connection.[17] That's good, even by today's standards. And the maximum bandwidth used was 200Mbits/s. It could be considered either a waste of bandwidth or that everyone behaved.

The main topic of the conference was political – the ongoing fight against the DMCA and similar anti-hacker legislation.[18]

Security and Privacy was another hot topic,[19] and the *"We are not living in a safe virtual world"* obvious conclusion was discussed thoroughly. This means that the risks have been known for 21 years now. But did companies evolve with the times? Far from it.

One could even say that, regarding this cyber war that has been ongoing for 20+ years, cybercriminals are the ones gaining momentum.

There were around 2900 hackers present at the conference, and this time I had a laptop. Yay.

One really interesting memory was a fish tank that was set aside for drowning mobile phones if they rang while a keynote presentation was being held. Amazingly enough, it remained empty.

And I still think it's something that should be done today.

2005 – WTH
(STILL, THE NETHERLANDS)

What The Hack 2005 was the last conference I attended.

Next to me, a "fellow hacker" was carding. To make it clear, carding is the trafficking and unauthorised use of credit cards.

This was way beyond the "hacker culture" I respected and grew up with. This was criminal.

I skipped *Hacking at Random 2009, Observe. Hack. Make. 2013, Still Hacking Anyway 2017,* and *May Contain Hackers 2022.* (2021 didn't happen because of Covid.)

In 2005, I got my first job as an InfoSec professional – White Hat.

As much as the '80s hacker culture had helped define me and my interests, the criminal activity that sprouted as a wild weed offspring was something I did not want to be involved with.

It's one thing to tinker and try to find new ways to circumvent software flaws, but it's a completely different universe when it's done for either criminal or political purposes, or just for a quick buck.

The hacks and breaches are now too many to mention. Plus, as big companies try to glue together the biggest number of netizens for their own profit and not to improve communication between humans (yes, I'm talking about Facebook and Twitter, along with many others), the risk and the reward also came up drastically.

We now see hacks on the news almost every single day. It has become, sadly, part of the norm.

Why? How? Let us try to dig a little bit deeper into that.

Throughout the years, I have worked in various industries: finance, banking, oil & gas, gaming, retail, insurance, fashion and travel.

Amazingly enough, the industry that takes cybersecurity most seriously (in my experience) is not finance; it's gaming.

The gaming industry is extremely regulated, and because the scrutiny is higher than in Finance, the fines and penalties are also heavier.

In the gaming industry, every single operation needs to be logged and ready for audit. That doesn't mean that the same is not expected in Finance, but companies take advantage of the leniency of the regulators.

I will use something that happened to me as an example.

Recently, while I was asleep and my mobile phone was quietly charging next to me, an operation was (hypothetically) made on one of my neobank accounts.

When I woke up, I saw that a premium subscription had been activated.

It wasn't by me.

I managed to cancel it and quickly got a refund. But the question remained – what happened?

My mobile wasn't hacked. (Really, I'm sure.)

So, was the bank hacked or exploited in any way? That's more likely than the phone next to me being breached.

I contacted the bank, who rudely told me, *"Yeah, the operation came from your phone."*

I asked them to *"Show me the logs from the server with the IP."* And they cheekily refused.

My IP. My data. And they told me *"No."* Apparently, they do not know what GDPR is.

Obviously, I made a formal complaint to the regulator, right after I moved all my money out of that account.

This corporate mentality of *"Let's make sure we don't hit the news for being bad at what we do"* just makes me realise that I feel a lot of Schadenfreude for some big companies.

Not all, of course. Many companies still think that they are too small to be the target of hackers or that *"It won't happen to us,"* so I have no ill feelings towards them. But, as a wise man once said, *"I pity the fool."*

One of the most talked about breaches in the 21st-century media was the TalkTalk UK breach.

Back in 2018, I and more than 300 (maybe 400) InfoSec

professionals was carefully listening to what Baroness Dido Harding (the TalkTalk CEO when it happened) had to say about the breach at the *InfoSec Europe Conference.*

There I was, front row, all ears.

We all listened carefully for an apology that never came. Of course, the company had already publicly apologised, but it was an HR template.

All we heard was a lot of *"Mistakes were made by someone else."* The crowd wasn't happy. I know I wasn't.

They had a vulnerability in some software for a long time, and I heard excuses about how they had done all due diligence, tech debt, and that the board was not tech-savvy.

We all know most boards are not tech-savvy. Yet, they are responsible and accountable when breaches happen. So, they need to understand their roles and responsibilities. This lack of *"They should have known"* was an attempt to diverge from the real issue – that C-level and the board have no idea about what's going on below them.

Stop. Let's talk about what boards and C-levels do to InfoSec professionals in the real world.

Real examples from my career:

- *"InfoSec tools are too expensive."*
- *"InfoSec is a big cost and brings no revenue."*
- *"We have to prioritise other things instead of fixing bugs."*
- *"You have been doing a great job, so we do not need to hire anyone for your team."*

Let's dwell on these for a minute.

Yes, most InfoSec tools are not free. Some are, but it's likely they won't cover all of your gaps and needs.

Conclusion #1 – tools and software are not free, and nor should they be. You can have a great team, but it won't be possible to have a good InfoSec mindset/culture and Risk Management based on human talent alone.

An InfoSec budget is a real necessity from the start.

Another common urban myth is *"We are too small to have InfoSec."*

No, you're not. No one is. Even if your company only has ten employees, if you do not have an InfoSec function or role, you're being naive, and I won't shed a tear if you're breached. It was your own wrongdoing.

The fact that InfoSec brings no revenue is another common and, quite frankly, ill-conceived excuse.

When you consider the cost of a breach (brand-wise, business-wise and even fine costs), when/if it happens to you, it becomes obvious that InfoSec is simply an investment – an investment in not getting breached.

Also, drop an infantryman/woman in a combat zone, remove their pay, food, supply, intelligence and communication and let me know how they do.

The prioritisation of resources is another pain point. It's widely known in the InfoSec industry that *"InfoSec will never trump product."* It has become more than a saying. It's what happens every single time, and even though it brings out a smirk from all InfoSec professionals, everyone knows it is impossible to change.

Unless, of course, a company is breached – and then, all of a sudden, budget is almost thrown at you, and everything you say is the most important task there is.

However, then it's too late – you have already failed to prioritise your resources.

The last example has to do with the InfoSec function and

team expansion. The rule here is, again, simple – the function/ team need to expand with the business.

How often can one see businesses hiring for 10/20 roles in Product/Engineering and only one for InfoSec (if that)?

Almost always, C-levels and the board still do not grasp the concept of InfoSec not only being a fundamental part of their business, but also part of the company's technical and operational foundations.

As I said, these are all real-world examples – sadly.

The internet is full of examples of things that were supposed to be good for everyone and make us all more secure. However, they ended up staining what cybersecurity is and making most people cringe when they see or hear the word.

Another demonstration of failure is the never-ending popups on almost every website, asking us to accept cookies.

What was supposed to be an improvement in data privacy is now nothing but a huge nuisance. Whatever companies decide to do with the consent and what information the cookies really gather is rarely audited and surely not enforced.

Let's face it; cookies are trackers. And websites make it really hard, or at least awkward, to say no to them.

So, billions of internet users are constantly giving permission to be tracked by a multitude of companies, tracking the herd's every click, every mouse movement.

Speaking of cookies: what about . . .?

THE CLOUD

There's a lot to talk about the Cloud and its intrinsic relation with black-hat hacking.

But let's start at the beginning.

First and foremost, the Cloud is an opportunity for many enlightened illusionary personalities to create . . . sorry – to make up dozens (hundreds?) of new acronyms.

I found out last month that DX stands for Digital Transformation and CX for Cloud Transformation. Not DT or CT.

Really?

As far as I know (AFAIK – see, I know my acronyms, too), the "X" was always used to depict eXperience – as in UX, most likely the first xX acronym invented. The thing is, it now means User Transformation because the X has a new meaning.

So how does eXperience become . . . tranXormation?

No idea. But then again, I do not have a TikTok account, so I must have missed the memo.

With DX now taken, I wonder what acronym will be created for Digital Experience. Maybe DIE. If books had emojis, this would be the place for a 'snicker.'

Moving on.

So, the initial sales pitch was: *"Ditch your mainframe(s), move to the Cloud, and your data will be safe while also saving you a lot of money."*

Wrong, wrong and wrong.

Simple debunking: you should never ditch anything until after your migration is complete.

Secondly, as with mainframes, you need cloud skills to keep your data safe, as the defaults for Cloud technology are – for the most part – not safe at all.

We have all heard of exposed data in the Cloud due to it being simply left unsecured, most likely because someone just did the classic "next, next, next, finish" and used the provider's default settings.

Infrastructure as code is solving some of these problems, but it's still not a turnkey solution.

As for saving money, it depends. It can happen, but it relies on a carefully planned Cloud migration with a mature roadmap; that does not always happen because many companies rush their Cloud journey.

Are mainframes expensive? Sure, they are. But a z14 can supposedly run up to 2 million Docker containers.[20] If you calculate the cost of doing the same thing in the Cloud, then it doesn't look so expensive anymore.

And, because these containers reside on the same machine, they can take advantage of co-location with other mainframe workloads and minimise the impacts of external network latency. As anybody who has worked with distributed architecture can tell you, a containerised microservice is only as good as its availability. Putting all containers on a single computer reduces the risk of interservice communication failure significantly. This is a very big deal – especially considering the 99.999% uptime of IBM Z. (And no, I'm not affiliated with IBM, nor am I endorsing their hardware – these are just facts.)

Slight pause to add a note: I am not against Cloud services. I had, and still have, some personal services running on four different cloud providers.

But one thing that rarely gets mentioned is that the Cloud represents, in and of itself, a plethora of new attack vectors that need to be taken into account from the moment a Cloud migration starts getting planned.

And this is a crucial aspect. When I say *"from the moment,"* anything that comes after is a huge risk and a breach waiting to happen. You can't expect to move to the Cloud and have criminals waiting for a gap analysis to be made, standing by while you fix your flaws. That simply won't happen. From the moment you have one asset exposed to the Wild West that is the internet, you will have people with nefarious intent knocking on your front door. So, it's better not to leave that door open.

Back in 2013, I made a presentation about the Secure Software Development Lifecycle for the company I was working for at the time. What really surprised me was the reaction of the audience to the first word, "*Secure.*" Everyone knew about the SDLC, but my focus on it being secure was seen as, to say the least, too ambitious.

And this is one of the engrained flaws in many tech companies. When security is only considered after a "*shift left*" movement that should be a default and not a tech hype, then the foundations are all wrong.

In that same presentation, I also joked that many acronyms would be invented in the future to cover the gaps in business processes and Risk Management, like DevSecOps, DevSecMLOps and DevSecMLAIOps. If only I knew then what I know now, I would have registered a couple of patents . . .

Another comment I made that I remember distinctly wasn't very well received was: "*There's a fine line between Agile and Fragile.*" Not that I don't believe that Agile can be good, but it shouldn't be a synonym for rushing things. "As soon as possible" does not mean not having time to consider security done right.

The audience wasn't amused, but the funny thing is that I still believe it, and almost ten years on, there's ample proof to back my (at the time) theory.

The obvious dad joke that can be used goes along the lines of *"Video Agile Killed the Information Security Radio Star."*

As much as I am trying to be blameless, I have seen it too many times not to mention it. Devs are under immense pressure to deliver the almost-impossible deadlines dictated by the product. Therefore, security was not involved nor considered, so there is not enough wiggle room in Sprint 324 to do threat modelling. And then there's no time to do any sort of pentest on the new shiny product because (surprise!) there are bugs that were caused due to the rush, and they will be fixed in Sprint 325. Rinse, repeat. Untackled/unsolved security issues are eventually abandoned, forever lost with a label of "Backlog." Then it becomes "Legacy." Then you're breached or ransomwared.

Surprised? You shouldn't be.

THE RISE OF RANSOMWARE
· ·

The top three ways previous ransomware breaches have entered organisations are phishing emails, email attachments, and users visiting malicious and compromised websites. While spam filters can prevent some of these phish from making it to the inbox and firewalls can block some of these websites, social engineering attacks now appear so genuine and realistic that more than a few will slip through the cracks. The primary barrier against such threats is the employees, and the strength of that barrier comes down to how discerning they are.

Additionally, many organisations work with Managed Service Providers (MSPs) or other third-party vendors that have access to their systems. If their security is breached, attackers may have a clear path straight into every business that MSP has as a client. This means that even if organisations do everything possible to make sure their own IT environment is secure, they also have to rely on the security of any third party that has access.

The desire to simply get it over with and pay the ransom to get data back quickly and return to business as usual is an instinct everyone can sympathise with. However, doing so is not a guarantee of recovery and also creates a vicious cycle.

Paying the ransom incentivises attackers to continue using ransomware. Even if you get your data back, giving in to demands only encourages further attacks on other organisations or even a repeat attack on your own. For example, the UK's National Cyber Security Centre (NCSC)[21] wrote about an attack on one company that paid £6.5 million to recover their data. Since the decryptor did restore their files, the company didn't investigate the origin point of the breach or its attack path. Less than two weeks later, the very same threat actor used the exact mechanism and ransomware as before to attack them again.

Further, paying the ransom encourages threat actors to increase their future ransom demands. In fact, according to the 2021 Cyber Threat Report,[22] the average ransom payment in the first quarter of 2019 was $12,762, while the average payment in the fourth quarter of 2020 was $154,108.

Finally, you simply cannot trust that attackers will return your data once you've paid. Experts almost universally advise not to pay the ransom. Once you've paid, they have what they want and face zero consequences for not holding up their end of the bargain. Despite this, according to the 2021 Cyber Threat report, 57% of organisations have paid the ransom. Unfortunately, 28% of these victims failed to recover their data. It's far better to invest the ransom payment into recovering the data through other means.

Many people think that once you receive the ransom note, the ransomware attack has begun. But in actuality, the note comes towards the end of an attack—once the data has already been encrypted. The median dwell time before the detection of ransomware is currently 24 days. This means attackers have all of that time to explore an organisation's environment, gain additional privileges, encrypt more data, or even steal sensitive information. However, the median dwell time before detection has gone down steeply – in 2020, it was 56 days. While this is certainly good news, unfortunately, attackers are quickly adapting, becoming increasingly efficient while remaining just as destructive. This narrows the window an organisation has between infection and extortion, making it more difficult to avoid the consequences.

Threat actors and cybercriminal organisations just recently demonstrated how quickly they could adapt during the Coronavirus pandemic. Taking advantage of the transition to remote work and general upheaval, ransomware attacks spiked in the first months. Phishing efforts increased dramatically, with Google reporting that they were blocking 18 million phishing emails a day that contained the keyword "COVID-19," plus 240 million emails with the simplified term "COVID."

Ultimately, attackers show no signs of slowing down their development of more frequent attacks and more sophisticated ransomware strains. According to a report[23] by the UK defence think tank, Royal United Services Institute (RUSI), ransomware operators are actively recruiting new people to improve their strategies and further advance the technology.

While the outlook may seem bleak, there are plenty of options to help safeguard your organisation. First, we must all have realistic expectations – ransomware breaches are no longer fully preventable. Instead, the goal is to put as many barriers as possible in place between an attacker and an organisation's critical, sensitive data.

PART V

BUSINESSES

THE CURRENT PROBLEM

We humans like to talk. A lot. Although, with the internet, it's more about typing than talking, to some extent, we all like to pretend everything we type is a speech to an audience of millions and that everyone agrees.

So, why is that a problem?

The problem is we talk a lot, but we communicate poorly.

InfoSec has always been entangled in different interests and priorities, and it is extremely challenging to juggle so many stakeholders and their roadmaps.

Real-world example:

Legal says security is a technical issue.

CTOs and CIOs say it is an employee issue.

Management says it is a technical and legal issue.

HR says security is *"trouble"* and *"Our job is to avoid trouble."*

Product team says they have many new features to implement that drive revenue, and the Security team does not.

They're also all busy with something else that should take precedence.

InfoSec should be the glue that unites them instead of being blamed for deviating from their roadmaps. Operations management, IT, HR and Legal need to understand that Security is a corporate responsibility, and not just something for the InfoSec team to achieve.

Easier said than done. InfoSec ends up being as much about people and stakeholder management as Tech and Risk.

Almost all InfoSec professionals I have ever known have gone through a period of stress, anxiety, anger, burnout, and even depression and generally poor mental health. A continuous stream of bad news about breach after breach can overwhelm any InfoSec practitioner.

This is a worrying statistic in a high-demand job where talent and skill are more important than ever to keep organisations safe and out of the cyber firing line.

Organisations remain as vulnerable as their security teams.

In an industry as tiny and tight-knit as cyber, acknowledging nontechnical problems is just as important as solving technical ones. You (usually) see an alert when your systems are breached, but you rarely get the same red flag when a colleague is struggling with a mental condition.

Security leaders shouldn't always be the ones to feel the blame when something goes wrong. In most cases, CISOs will have requested budgets, assets, and changes that weren't signed off – so they must be ready to remind the board that security is a shared responsibility.

When addressing mental health, security researchers argue that cybersecurity professionals and criminal hackers differ primarily in their state of mind. They perform similar tasks, such as manipulating their target – the only difference is their goals and what they choose to achieve them.

Empathy plays a huge role in determining the good versus the bad, and what if the practice of empathy could be supported through an organisation's desire to boost mental health?

If security teams are looked after, they're less likely to fall into states of stress, which, in turn, could avert erratic urges, such as paying a ransom without consulting leadership or revenge hacking. In extreme circumstances, it could prevent security professionals from "going over to the dark side" to utilise those same skills they used to protect a company but for a detrimental cause – executing a cyberattack instead.

Supporting mental health fuels a team's best work and

keeps them in alignment with their core values, which only maximises sufficient protection for the organisation itself.

We can't instantly offer security individuals the downtime they need to refresh, and we certainly can't change the way criminals strike. However, we can change our attitude toward mental health and encourage others to speak up and ask for help.

THE TECHNICAL DEBT

Tech debt has become the norm – every company has it, even the ones that don't admit it. And "it's ok."

No, it's not ok.

Also, it's not a debt – allow me to expand on this.

By definition, tech debt is the result of prioritising speedy delivery over perfect code. The real-world example is Product trumping Security because of the need for revenue. Features get fast-tracked (as in, corners are cut) without going through Security due diligence because of fear it might delay the implementation of a feature that wasn't discussed with Security in the first place.

But a debt is something that can be repaid in one go. If, for example, one wins the lottery or receives an inheritance.

Tech debt goes beyond that. So, the only metaphor that can be applied here is pollution: something created a long time ago that doesn't get fixed fast, and even though you're aware of it, there's no turnkey solution for it.

Thirty years of underinvestment by the industry, 30 years of profiteering, 30 years of regulatory failure, and 30 years of a vacuum of political oversight.

THE SKILLS GAP. NOT.

Another important urban myth is that *"There is a cybersecurity skills gap."*

There isn't.

But the job market does have a lot of problems under the surface that drive InfoSec professionals away.

The one I am putting at the top is job specs.

I have seen too many job specs where the "requirements" and "expectations" are two pages long, including ten years of mandatory experience for an entry-level InfoSec role.

It was most likely copy-pasted from somewhere else, and the spec has nothing to do with the company objectives. Maybe the company does not know what their objectives should be – that is fine. But don't expect a myriad of applicants. The good professionals will have read your poor spec and moved on. So, the applicants you have are professionals who are sort of desperate and just want "a job."

And then you blame the industry and the cybersecurity skills gap.

Another flaw in job specs is the lack of salary expectations, or at least a salary range that is fair for the role you're hiring for. I have no words to express my deep hatred for the phrase *"Depends on experience."* If that were true, I would skip nine out of ten job specs because, with my experience, I sit at the top end, and it's very likely that by not giving me a number, you can't afford me.

We all know companies will try to recruit an InfoSec professional for the least amount of money. That is also wrong. A company should have a budget for their requirements – not more, not less.

Be honest about it, put your money where your mouth is, and don't blame it on the cybersecurity skills gap.

Another pet peeve of mine is the *"Benefits – holiday pay and pension,"*

There are cases where this is true – final salary schemes, known as 'Defined Benefit' schemes. The pension paid out by these schemes is defined as a 'benefit.' But in every other situation, it's not a benefit, and I just ignored the job spec you posted. If a company is trying to evade a worker's rightful pay from the start, then it's not a good place for anyone.

Here's a real benefit: "Unlimited holidays."

If a company wants to slam an acronym that will attract good talent (and this is not specific to InfoSec), they should mention they are Equal Opportunity Employers.

Another common finding in job specs that make Security professionals run away is: *"Needs to be a respected/liked member of the InfoSec Community."*

What?

Will companies really look at the number of followers on Twitter or contacts on LinkedIn to ascertain the true value of a professional?

That is not just wrong; it's also insulting.

I have met many InfoSec professionals who didn't become my best buddies, but they were truly amazing at their job. That is just human nature. You can't be friends with everyone, and you are not flawed if that is the case. The fact that companies mention this in a job spec indicates how little time is spent by whoever creates said specs, without any empathy or compassion for everyone whose number of followers does not define professional abilities.

Skills gap? No.

What does exist, is an opportunity gap. Job specs asking

for some experience are not a bad thing, in essence, but the lack of entry-level InfoSec jobs is staggering. That gap is real.

By widening the net and encouraging more neuro-diverse talent into the IT/InfoSec/SOC area, employers can help to alleviate hiring challenges. More thought should also be put into making InfoSec a career destination in its own right, rather than a jumping-off point. That will help encourage greater retention and create a blend of experienced InfoSec professionals and new blood.

Organisations need to think about cultural change. Security is still viewed myopically as the sole responsibility of the CISO. Yet what happens if a board fails to sign off on new tools or process changes per the CISO's request, leading to a breach? Who is responsible then?

The reality in this situation is that the board themselves are accepting the risks outlined by the CISO and are ultimately accountable for any breach as a result of failing to invest. The truth is that every staff member across the organisation should come to see themselves as a quasi-security professional – invested in the benefits of getting security right and aware of the dangers of doing it poorly.

This would elevate the role of the security function within the organisation and, perhaps in time, lead to earlier engagement in business initiatives. When security is addressed in projects early on, it minimises the chances of reactive firefighting later down the line.

I have recruited for several positions in the past, and I spent a lot of time creating appealing job specs. No job spec will ever reveal the true depth of the role, but the fundamentals need to be there. *"Why would you want to work for us? Here is why"* should be the focus, and not *"We're leaders; everyone wants to work for us."*

With that blunder out of the way, we reach the face-2-face stage. Nowadays, this can be done remotely, as COVID forcefully taught us all.

In this stage, there will be obvious questions about past experience, or how a candidate successfully managed to get their point across over some adversity from the business.

But that's not what I am looking for when I am a hiring manager.

I am looking for a curious mind. Someone who wants to do better every single day and understands that InfoSec is, and always will be, challenging. Someone who is not afraid to experiment and challenge the Status Quo.

Not afraid to challenge ME.

I think it's clear by now where I'm going.

Yes, I am looking for a hacker mindset. Not a Red Hat, not a White Hat, not a Purple Hat, nor any other hat colour.

Just . . . a curious mind.

WHY DID YOU LEAVE?

Last but not least, in my list of things that make Security professionals run away: *"So, why did you leave your last job?"*

Every single time this question is asked, I think I made a mistake by applying to that role.

The real answer is: *"It's none of your business, and it shouldn't influence my application."*

It might be because I relocated due to a too-long commute that became untenable. Maybe my contract was over. Maybe I couldn't relate to the company's values or goals. Maybe I didn't get a raise after three years. Maybe mind your own business.

This question is truly personal, and it shouldn't be part of a first contact. If the reason why I left my previous company is a deciding factor, then you are already losing me as a candidate.

It almost feels as if companies are forcing impostor syndrome on candidates and then blaming a "skills gap."

I'm happy if I'm asked about my salary expectations (note: not how much my salary was at my previous job, that's also out of your scope), but questions like these should only come in a friendly environment down the line, possibly close to the *"So, do you prefer cider or beer?"* question.

NATION-STATE THREATS

Breaches, data leaks, and hacks of various and assorted natures can be seen on the news almost every single day.

And those are only the ones we know about, as many will only be revealed to the general public a long time after. If they ever are.

This widespread apathy and lack of transparency for InfoSec Risk and Incident Management have paved the path for corporate espionage, theft and many other kinds of wrongdoing.

In the same way that some more mature companies use a matrix of Probability vs Impact to measure their Risk and Investment, criminals also have a non-spoken matrix on their side: Effort/Risk vs Reward.

The analogy I keep using is, *"Would you leave your house door open and hope no one would steal from you?"*

To the criminal, these "open doors" serve as an opportunity. Even if they are low-risk and offer an unknown reward, it's still profitable. And that's what it is all about nowadays – reward.

Marriott has had seven (that we know of) data breaches since 2010. Maybe the second time was bad luck. Maybe the third time was because they had no time to do a Risk Assessment. Maybe the fourth was regarding a new attack vector they didn't know about.

But seven times?

Come on.

The latest one was downplayed as *"Non-sensitive internal business files and was a result of social engineering at one single hotel."*

So, what, Marriott? Are you trying to tell us, "That's ok?" Or that it's "expected?"

No. That is wrong.

The truth is, most companies don't care about the privacy or security of your data. They care about having to explain to their customers that their data was stolen.

Every industry has long neglected security. Most of the breaches today are the result of shoddy security over years or sometimes decades, coming back to haunt them. Nowadays, every company has to be a security company, whether it's a bank, a toymaker or a single app developer.

The threat landscape continues to grow in the volume of attacks that occur daily and the variety of methods used by cybercriminals. Attacks are coming at a ferocious pace, and a single data breach could cost a company millions of dollars along with massive amounts of time. Of course, the ultimate threat is a ruined reputation that can damage the business for years to come.

Back in 2016, the world was introduced to the Petya encrypting malware, and the awareness changed. Nowadays, everyone knows about ransomware.

NotPetya followed in 2017, and since then, there has been

a flurry of ransomware variants, and there are no signs to indicate the impact or occurrences are decreasing – on the contrary.

So, what changed? Meet nation-state actors.

A nation-state actor has a "Licence to Hack." They work for a government in order to disrupt or compromise target governments, infrastructure, organisations or individuals to gain access to valuable data or intelligence, and can create incidents that have international significance.[24]

Personally, I think calling these criminals "actors" is an insult to real actors. Just because these criminals are state-sponsored, it should not give them any sort of positive protagonism that misrepresents what they are doing – a criminal activity.

The sad choice of words came from NIST's attempt to standardise the terminology used to describe cybercrime in NIST SP 800-150 (Guide to Cyber Threat Information Sharing).

NIST is the National Institute of Standards and Technology, a physical sciences laboratory and a non-regulatory agency of the United States Department of Commerce.

Unlike other types of cybercriminals, who exploit a vulnerability and move on, nation-state attackers are persistent and determined to achieve their objectives. They invest serious time profiling their targets, probing their network for vulnerabilities, and continually adding more tools and skills to their capabilities.

Data theft is typically the purpose behind cybercrime, but that's not the only goal of these criminals, or even the most common. Instead of just snatching data, nation-state hackers like to take it a step further. They use tools like ransomware and other malware to shut down manufacturing, interfere with logistics, and disrupt important research.

When news[25] broke that hackers working on behalf of

a Chinese intelligence agency may be responsible for the Marriott breach, questions abounded. Why would China be interested in loyalty program data by the millions? And why hospitality data?

Foreign intelligence agency actors aren't exactly interested in earning a free night's stay at a Marriott property. The answer is potentially far more nefarious. The fact is that data collected from breaches are but one piece of a larger, darker puzzle. Stolen customer data – when combined with travel data (see Delta, Cathay Pacific, and British Airways hacks, among others) and other sources of online personal information (i.e., what we share across social media platforms) – enable intelligence agencies to build profiles on individuals. These profiles can then be leveraged to recruit potential informants, as well as check the travel of known government and intelligence officers against their own government to identify moles.

It's also critical to note that heads of state and other political VIPs are no longer foreign intelligence agencies' only marks. Ordinary citizens are similarly targeted, especially those who may have unfettered access to troves of company intellectual property (IP) that a foreign government may want for their domestic economy.

In the case of large enterprises, CISOs and C-suite executives are focused on individual pieces of lost data versus the sum of what that data can reveal about an individual as a whole, putting them (and us) at a significant disadvantage. Indeed, the entirety of the digital footprint we create, which can be used to impersonate us or to profile/create leverage on us, is greater than the sum of the individual data parts. Likewise, consumers don't typically consider the bigger picture their personal data paints regarding their travel patterns, purchasing habits, hobbies, (not so) hidden secrets, social causes and more. Add in breach burnout, wherein the public has become desensitised to countless stories of data exposure, and a perfect storm for harvesting operatives and stealing IP emerges.

Until enterprises view data holistically and realise that any company with valuable IP could be the target of a foreign government on behalf of that company's foreign competitors, they will continue to play into the hands of transnational threat actors at the expense of consumer safety and national security.

It is critical that organisations incorporate cybersecurity into the very fabric of the business, from the C-level down. This includes training and education, as well as seeking expertise from security service companies who understand how to protect organisations from the capabilities of foreign intelligence groups. That education must include an understanding of how personal, government and business-related information can be used by foreign intelligence agencies, plus how corporate IP may be of value to foreign competitors. Whether it's a game of chess or an intricate puzzle, individuals must look beyond the breach at hand and grasp what's around the corner.

Don't forget that state-sponsored threat actors may be politically motivated. As such, their goals for the attack are not always clear and can change over time, compared to threat actors purely motivated by cyber-theft for money. This can lead to an unpredictable and challenging legal response.

Dmitri Alperovitch, co-founder and former chief technology officer of CrowdStrike, believes the Russian government may target Western organisations in retaliation for sanctions recently imposed by the US and other governments as part of the ongoing Russia-Ukraine conflict.[26]

Put another way, the Russia-Ukraine conflict has changed the landscape and motivations. It may motivate Russia to try to hurt Western organisations, steal data or spread ransomware in an attempt to recoup lost money from sanctions.

Many IT security leaders express a high degree of confidence in their ability to not only defend against but also trace the source of an attack accurately. However, the data reveals the hubris: there have been 5.1 billion breached records and 1243 security incidents (that we know of) in 2021.[27]

There has been an increase of 11% in security incidents from 2020 to 2021, but the number of records breached is only 25%. That does not sound right.

It is now becoming clear that companies both fear and understand the brand (and, many times, also legal) impact and implications after a breach. Therefore, there is a fair degree of certainty that not all companies reveal their breaches nor the correct number of records that have been compromised. Is it shame? Fear of fines? Without any way to know for sure, my gut feeling is "both."

In a world governed by nation-states that is increasingly reliant on digital systems, it is no surprise that national security is correspondingly becoming synonymous with cybersecurity. Cyberspace has now been established as the fifth domain, following land, sea, air and space.

As for nation-state conflict, states are using cyber operations as a low-cost tool of statecraft to achieve strategic objectives unless they face clear repercussions. These objectives are political, economic, and military advantages.

These cyberattacks are not arbitrary, one-off, meaningless acts of state aggression. They have distinctive characteristics, which include everything from motivation to target to the type of attack, and they are not an end in themselves.

Because of the severe lack of international norms and agreement concerning cyberspace, targeted cyber operations never meet the criteria of an act of war or aggression. However, if there were norms in place, perhaps some of these advanced and impactful attacks would no longer fall into a grey area below the threshold of total war.

This gap in international frameworks is an exploitable uncertainty and one that is worth attention, especially in light of the threat posed to critical infrastructure. International and domestic frameworks that apply to traditional warfare are outdated and irrelevant when acts of aggression from the cyber domain come into play. Long-existing theories cannot be readily imported into handling nation-state cyber operations. At the very minimum, our institutions and norms must be applicable to this undeniably emerging pattern of warfare.

Rolling out new IoT devices and software, plus expanding the number of people who access mobile and internet devices regularly, only increases the potential for new system vulnerabilities and their exploits by criminals. In addition, the speed at which this domain changes surpasses the realistic amount of time for countries and international organisational bodies to negotiate and legislate.

In his book, War in Cyberspace, former US government official and counterterrorism expert Richard Clarke asserts that *"Cyber war is a wholly new form of combat, the implications of which we do not yet fully understand."*[28]

RECYCLING

. . . .

The phenomenon of malware recycling is unique to cyber conflict and encompasses both the permeation and fluidity aspects of the tools used. When a nation-state launches a missile or engages in battleground warfare with an opposing state, those strategies and tools aren't repurposed for later use by other states or actors. In state-level cyber operations, however, malware, ransomware and viruses are often deconstructed by various entities (security researchers and firms along with the government). Pieces of the code are dissected, whether for understanding the adversary's capabilities or for patching vulnerabilities. Different segments of malware can be sold and used again, as has been seen in even nation-state cyber conflict. The US-Israeli-made Stuxnet worm has been repurposed for acting on vastly different objectives.

A 2013 NSA document leaked by Snowden indicated concern and evidence that Iran demonstrated replicated techniques from US malware such as Stuxnet, Flame, and Duqu in their well-known *Shamoon* malware.[29]

Shamoon is also an example of an attack that has been modified by its creators to strike again at different targets. According to firm FireEye, the 2018 *"Shamoon 2.0 [was] a reworked and updated version of the malware [they] saw in the 2012 incident."*[30]

This challenge, unique to cyberwarfare, is one of the many new challenges that has emerged with the rise of cyber as the fifth domain of warfare.

Participation and attribution contribute to the opaqueness of cyber operations as well. Non-state actors with sophisticated technical skills (and at times hired to execute governments' cyber operations) blur the lines between state cyber operations and non-state cybercrime. This presents a unique challenge for nation-states combatting and responding to this type of attack. Specific to the problem of attribution, while identifying the actor behind malware campaigns and cyberattacks, therein lies a unique challenge of its own. "False flagging" is also employed by sophisticated state APT actors in order to deceive and mislead victims and the security community from correctly identifying the origin or identity of the attack. As Russia publicly denied any involvement during their DNC and US election hacking campaign, they intentionally disguised their identity with a false persona of Romanian origin.[31]

And what if a state defends itself from attribution by placing the blame on "non-state actors" who happen to have operated within its borders? Should the law attribute the malicious activity of non-state hackers to the state?

This is a particular problem for the law of attribution and cybersecurity, given that the relatively low cost of conducting a cyber-attack opens the option up to myriad non-state actors acting for a variety of motivations. Additionally, all of the typical problems associated with attributing an attack risk further attenuation between the individual conducting the hack and any chain of command or control infrastructure tying that actor to a state. After all, hackers do not wear uniforms in cyberspace. Thus, a law of attribution must address the inevitable result where it follows the trail leading to an individual hacker and faces the problem of how to connect that person to a state for the purposes of legal responsibility.

Suppose a state has suffered a cyber-attack and wishes to bring a legal claim attributing that attack to another state. With everything laid out so far, the state knows the procedure for initiating action and the back-and-forth sequencing of complaint and answer, summary judgment arguments, and the production of the evidence.

Here, in this last step, the state runs into a problem: what happens if significant portions of the evidence on which it relies are derived from covert intelligence?

Moreover, states may have plausible factual bases for attributing an attack. However, they may not want to disclose such evidence on legitimate grounds since cyber-attackers could learn from those points of attribution and avoid leaving the same fingerprints in the future.

The law of attribution faces the challenge of reconciling the need to present such evidence with states' desires to preserve the secrecy of their confidential intelligence and sources.

Two or three decades ago, *"What happened in cyberspace stayed in cyberspace."* Nowadays, not really.

Accountability goes hand-in-hand with the challenge of identifying and responding to cyber conflict in different ways than with conventional warfare.

Naturally, as correct attribution is difficult already, it is especially difficult to hold states accountable for malware campaigns they were only allegedly behind. It is difficult to be held responsible for cyber activity in the same way one may be for developing or testing nuclear weapons, for example, as this approach to warfare does not require huge physical development facilities or physically detectable testing.

We now know that nation-states have appropriated and adopted hacking techniques and knowledge as a tool of war. Without fear of facing repercussions and being held accountable by the global community and/or established international institutions, there seems to be little that prevents a nation from developing and using these capabilities, especially if they can get away with it not being attributed.

The nature of certain cyber operations, such as persistent engagement, network penetration, probing and reconnaissance (as mentioned above when examining the temporal and physical aspects of cyber conflict), also present a challenge while distinguishing benign, defensive cyber activity from that of reconnaissance efforts in preparation for a targeted malicious cyberattack. Active defence measures or monitoring can look very similar to conducting reconnaissance. It raises questions such as whether or not reconnaissance efforts should be treated like an attack, or how to respond to Russian or US electrical grid probing and access to networks. Misinterpretation of another state's activities in cyberspace, or misattribution of attacks or operations could very possibly lead to escalation, especially, for example, if a malicious or self-interested actor were to route attacks on the other side through US or Chinese servers during a tense period in the bilateral relationship.

These potential conflicts blur wartime and peacetime and elucidate the gaps (or even irrelevancy) in traditional definitions of offensive and defensive state behaviour.

The idea of using force to prevent or stop crime is intuitive in the physical world. You can fight back against an attacker. You can tackle a purse-snatcher. You can reach into the pockets of a shoplifter before he leaves your store. You can hire rough men – even armed ones – to guard your belongings. Of course, there are many things that you cannot do, and reasonable people can disagree about the limits of these actions. However, the law generally recognises that force is sometimes necessary to defend persons and property, halt ongoing crimes, and prevent suspects from fleeing.

The rights of private entities to use reasonable force have not extended to cyberspace. Under current law, it is illegal for the victim of a cyberattack to "hack-back" – to launch a counterattack aimed at disabling or collecting evidence against the perpetrator. This blanket prohibition imposes enormous constraints on the private sector's ability to respond to cyberattacks.

Criminalising self-defence outright would seem ridiculous in the physical world, but cyberspace blurs the traditional conceptions of property, security, self-defence, and the role of the state. Consider these questions: Is intruding onto my computer the same as intruding into my home? In both instances, my property and privacy are being violated. Can I pursue a cybercriminal through the web the same way I would chase a purse-snatcher down a busy street?

Legalising counter-hacking would allow companies to carry out their own vigilante justice against the accused with no due process of law. Private companies may launch attacks indiscriminately with little evidence, or they may inflict far disproportionate punishment on an attacker.

There's also the fact that innocent third parties may be harmed in counterattacks. Often, cyber threat actors will hijack unwitting victims' computers to carry out an attack. These computers could become collateral damage of a hack-back. Legalised hack-back could have international implications if a private company finds itself attacking a nation-state actor. This would be dangerous, not only because the nation-state would likely far outmatch the private company, but also because the fight could escalate and become an international conflict.

The problem with these assumptions – and current law – is that they do not distinguish between different kinds of counter-hacks. To be sure, counter-hacking has the potential to infringe on the privacy and property rights of criminals and third parties. However, hack-back techniques have varying effects that may be appropriate in different contexts. These techniques can be categorised on a spectrum of "utility" based on the severity versus the benefit of the counter-hack. Severity refers to how destructive or invasive the counter-hack is. Benefit relates to how effectively a technique accomplishes some legitimate purpose, namely stopping an ongoing attack, protecting data, or gathering evidence.

IT'S NOT ALL ABOUT RANSOMWARE
· ·

With ransomware holding steady as one of the most significant threats facing businesses and individuals today, it is no surprise that attacks are becoming increasingly sophisticated, more challenging to prevent, and more damaging to their victims.

The endless stream of cheap – or even free – data has a value that criminals seek, both for its value and simplicity to get ahold of.

But this information is not a secret.

So, why do ransomware attacks happen on a daily basis?

Mainly due to delusions of grandeur like *"We're secure enough," "It won't happen to us,"* and the quintessential *"We'll never get breached, so we don't need to train for it."*

As Amar Singh, an InfoSec expert, so aptly put it:[32]

· · · · · · · · · · · ·

Why do organisations get caught with their pants down during a Cyber-Attack?

Stop being reckless. Most companies operate in Cyberspace without a robust & tested IR plan for a cyber-attack. This is akin to living in an earthquake-prone area (think Japan) without planning their response. Or going rafting without a life jacket! You get the idea.

Consequently, the journey to a cyber resilient business must start with senior leadership acknowledging that cyber-attacks are a reality and that their business will succumb to an attack in cyberspace; they could be attacked by the averagely skilled cyber-attacker; a determined attacker will succeed in defeating all defences, no matter the budget & technology.

Focus on planning, planning & more planning. Plan on being able to detect various types of intrusions and attacks on your critical assets; how you will respond when the attacker succeeds in breaching your defences; how you will recover and resume your business operations; ensure management understands their role during and after a cyber-attack; have incident response playbooks for management, communications and technical teams; conduct regular tabletop exercises to build 'muscle memory' – this also reduces the '1st-time-in-an-incident' panic; prepare and practice over and over again.

Have easy to read, easy to find IR & Comms Playbooks – and these must be easy to read and follow during an attack, and fit for purpose.

Trust AND doubt the technology stack: stop putting complete faith in tech and run from those who promise the world (100% detection rates); assume failure, train the human and trust their gut. Your biggest liability can also be your greatest strength.

Resources: Hire trustworthy people and partner with a company who can support you throughout your journey.

. . . .

I agree with everything said above. Easier said than done, though.

ABOUT DATA

As a consumer society, we cannot rely on big companies to secure our data.

Can they do it? They can at least try.

But do they?

As an example: Meta's "Responsible Innovation Team," a group meant to address *"Potential harms to society"* caused by Facebook's products, was disbanded.[33]

Why does that matter?

Well, for Meta, it surely does not matter. They did not explain why this happened (they don't have to, and most likely they will never do), and yet their stock price was up 2.39% the day after it happened.

And this casts a gloomy shroud of darkness on the current state of data security for us common mortals. If companies like Meta disband a team that was supposed to defend us, consumers, from potential harm caused by its products, who is doing it for us?

Nobody.

Ultimately, it falls on each and every one of us to follow industry best practices where our data is concerned.

But how many of us, realistically, are in a position to do it? A very small percentage.

And who will keep taking advantage of that humongous gap? State nations, big companies, careless companies (regardless of size), and every malicious individual who has access to the opportunities that the online jungle presents.

We've all seen it when accepting a cookie somewhere: *"We may collect consumer data and use it to power better customer experiences and marketing strategies."*

I think the "may" is deeply hilarious. What they really mean is, *"We will use and abuse this data and sell it for revenue in any way we can."*

As technologies that capture and analyse data proliferate, so do businesses' abilities to contextualise data and draw new insights from it. Through consumer behaviour and predictive analytics, companies regularly capture, store and analyse large amounts of quantitative and qualitative data on their consumer base every day. Some companies have built an entire business model around consumer data, whether they sell personal information to a third party or create targeted ads to promote their products and services.

Businesses that are so far untouched (or simply unbothered) by data privacy regulations can expect a greater legal obligation to protect consumers' data as more consumers demand privacy rights. However, data collection by private companies, is unlikely to go away; it will merely change in form as businesses adapt to new laws and regulations.

Other than privacy implications, automated data collection and processing may have a secondary negative impact: discrimination. Automated data mining is used in several services to derive association and classification rules, which are then applied to a variety of decisions, such as loan granting, personnel selection, insurance premium computation, etc. While an automated classifier may be seen as a fair decision-making tool, if the training data are inherently biased, the generated rules will result in potentially discriminatory decisions.[34]

Let's take another small pause to retrospect on the initial meaning of the word hacker and the stories I have shared with you, the reader.

What is happening today is not what hacking meant 40 years ago. These criminals do not represent the hacker culture.

And this is really not what *Hackers*, the movie, was about. That's my main reason for truly hating the "actors" label.

If you ask me if any of these nation-state employees attending Hacking In Progress 1997 were sitting next to me, I am almost sure the answer is *"No."*

These criminals represent a new generation – surely not the *"beauty of the baud."* Nations now train and harbour these individuals, but they are nothing but e-criminals. (I think that might be a new word.)

I'm not saying there are no 50-60-year-olds that are criminals. I've met some shady figures in the past. But they are an exception, not the rule.

And sadly, as the word "hacker" evolved, the perception of the word changed, too. What was once almost a compliment has now become a synonym for criminal.

The word will keep evolving, I am sure. But a lot has been lost since it was created. We, as a society, have progressed immensely with so many things – especially anything internet-related. However, we have lost what I consider a treasure, and we risk it eventually being forgotten: the original hacker culture.

PART VI

WHAT NEXT?

Spoiler alert – *things will not get better overnight, nor easier.*

FROM HEROES TO HOOLIGANS

Nowadays, when we hear about hackers, it is usually as anti-social, possibly dangerous individuals who attack systems, damage other people's computers, compromise the integrity of stored information, create and distribute viruses and other harmful code, invade privacy and even threaten national security.

They flout the law by cracking into communications networks, copying and distributing copyrighted software and other intellectual works, caring nothing for the norms of common morality. They stay up all night and take on strange and menacing names like Legion of Doom, Scorpion, Acid Freak, The Knights of Shadow, Terminus, Cult of the Dead Cow, and The Marauder. To top it off, the essential credo of old-style hackerdom – creative brilliance above all – has given way to a culture of "script kiddies" or "copycats," who merely mimic the technical ingenuity of a few creative hackers in order to further anti-social and often selfish ends.

What accounts for the transformation in our conception of hackers from *"heroes of the computer revolution"* to white-collar criminals and terrorists of the Information Age? One straightforward speculation is that hackers themselves have changed. They no longer discriminate in their target choice; they victimise not only centralised bureaucracies, carefully chosen for their obstruction of the "hacker ethic," but also unsuspecting users and consumers of the digital media. Having cut themselves adrift from their idealistic moorings, they are no better than other common criminals, intruders, vandals, and thieves. We see them as villains now because now they are villains.

Another speculation points not to a change in hackers themselves but, largely, in us. Because our standards and values have changed, what we used to admire or tolerate, we now deplore. Value shifts such as these are not unprecedented; consider the cases – more significant, obviously – of slavery, racism, sexism, etc.

These suggestions hold some truth, but they form a dualism that begs for synthesis. My own account seeks such middle ground by reading the transformation against the backdrop of a shifting social context. However, before considering this account, we should review two others that have drawn contextual phenomena into their stories.

One, offered by Deborah Halbert,[35] hypothesises that the shift in our evaluation of hackers results from a conscious movement by mainstream voices of governmental and private authority to demonise and portray hackers as abnormal, deviant bullies who victimise the rest of ordered society. Hackers are presented as the new enemy of the Information Age, an age in which old enemies (for example, the Soviet Union) have dissipated, and the world order has shifted. Mainstream media, law and government focus on the destructive acts of hacking in an effort to construct a new enemy and to justify systematic lines of action, such as very public indictments of particular hackers (e.g., Kevin Mitnick, Robert Morris).

Demonising hackers serves two ends that are important to the government and established private powers. The first is to control the definition of normalcy in the new world order of computer-mediated action and transaction; the good citizen is everything that the hacker is not. According to Halbert:

- - - - - - - - - - -

"It is the role of the deviant to mark the boundaries of legitimate behaviour. Hackers, constructed as deviants, help [to] define appropriate behaviour and appropriate identities for all American citizens, especially in a computer age where ethical guidelines are still ambiguous."

- - - - - - - - - - -

The second is the justification of further expenditures in security, vigilance, and punishment. To the extent that established powers can persuade us of the severity and urgency of hacker threats, they are likely to elicit support for security measures, including governmental vigilance over the internet, greater financial investment in safeguarding computer systems and information, and tougher sanctions on hackers.

In a similar vein, in 1991, Andrew Ross[36] portrays the changing moral status of hackers as a cultural regrouping, with hackers pitted against the corporate and government mainstream. He suggests that, in entrenching the association between hackers and viruses, mainstream culture linked the hacker counterculture with sickness and disease, particularly with such stigmatised diseases as AIDS. According to Ross, making this link helped mainstream forces to generate equivalent hysteria in the casual user and moral indignation in the legislature. At the same time, software vendors benefited from public distrust of unauthorised copies of computer programs.

In the process, *"a deviant social class or group has been defined and categorised as "enemies of the state" in order to help rationalise a general law-and-order clampdown on free and open information exchange."*

As with the explanations proposed above, my own account brings various contextual and historical factors to the foreground. However, it does so from a different vantage point, with greater specificity and the benefit of a larger temporal arc and first-person experience.

My main thought is that changes in the popular conception of hacking have as much to do with changes in specific background conditions, changes in the meaning and status of the new digital media, and the powerful interests vested in them as with hacking itself.

It is a known fact that 83% of statistics are made up.

Or, as Mark Twain popularised the quote most attributed to the Prime Minister of Great Britain, Benjamin Disraeli, *"There are three kinds of lies: lies, damn lies and statistics."*

But as much as companies keep looking at the rising number of businesses that are hacked daily, the number is still going up.

However, institutionalising a scare culture will not help.

Some so-called expert reports refer to "ransomware attacks," which are just attacks that have not been successful. Yet the whole narrative cultivates fear[37] (which, to some extent, is warranted) but without clear indicators of which type of attack is being enacted.

Statements and statistics like *"half of organisations (61% US and 44% UK) have been the victim of a successful ransomware attack in the last 18 months"* can, in fact, have very serious and adverse consequences. Companies might feel, *"it's happening to everyone, so there's nothing we can do anyway."*

Does a phishing email represent a cyberattack? Yes, it kind of does. Does it represent a ransomware attack? Maybe. Maybe not.

Did I mention 83% of statistics are made up? I meant 84%; the number has just increased.

My personal RSS feed (yes, dear reader, it's clear by now

that I am old-school) was carefully selected in order to include only respectable and reliable sources of technical news, mostly InfoSec-related.

However, looking at my feed just now, the current news stories are:

· · · · · · · · · · · ·

Mark Zuckerberg's avatar, a walrus that was euthanised (RIP), Trump's Truth Social and some Pokémon stuff.

· · · · · · · · · · · ·

What does that tell me? That my "faithful" sources are all part of the problem – there is a very blatant intention to keep the news coming, and if there aren't any, then just inflate the importance of others. Seriously, why do we have time to even discuss something as trivial and banal as that, and why does it have so many comments?

Because we no longer filter. We are happy to ride the data wave, and any comment, as dumb as it may be, gives us the impression that we're participating.

It reminds me of some Amazon comments I saw on the Customer Questions & Answers section, where many people answer "I don't know" to a question instead of (logically) not saying anything.

Seeing as the only two RSS feeds worth my time are *Angry Metal Guy* and *The Onion*, something is off.

As data becomes more and more valuable (and I'm referring to consumer data, not the mindless stream of data that social media provides us), criminals will keep devoting more and more of their time to companies that hold the goods.

Users are now more aware of data breaches impacting big companies. Yet the amount of people who still do not use Two

Factor Authentication or use a stupid password that, according to them, "is pretty good," even though criminals take minutes/hours to crack them, is still on the rise, and it will become the trend.

One thing that has grown, along with cyber threats, is the herd mentality of *"it won't happen to me"* or *"no one wants my data."*

This is wrong, but mentalities are hard to change. And even if they weren't, the big conglomerates have no use for a herd that *"Does the right thing."* They depend on the mindless click, on the impulse buy, and for that to keep happening, they only need to let the herd . . . well, behave like a herd.

Of course, cybersecurity vendors will also keep evolving and creating more and better products. But the criminals will evolve, too.

With the introduction of the Internet of Things, like smart cameras, smart washers and digital home assistants, we were all told that these would make life more convenient, but they are also notoriously prone to hacking.

We're selling – and most of the time, giving it out for free – our data (and ultimately, compromising our freedom) because we want to have the right to be lazy.

I sometimes wish connecting to the internet would become a niche and complex activity. Not trying to compromise my fellow person's rights, but I just preferred it that way.

As of right now, regulators are the quintessential hangman who does not know what a noose is. I remember the BA 2018 UK breach, where the other 420,000 victims (customers) and I were initially told that we were entitled to compensation.

Sixteen thousand compensation claims were submitted after going through too many bureaucratic hoops. I submitted a claim but never heard back from them. Not even a *"We do not care about you, as you are not a Flying World Traveller Plus card holder."*

The initial fine of £183m for GDPR infringements was reduced to £20m after the regulator took into account the impact of Covid-19 on BA.

The reduction was laughable, as the regulator colluded with the lack of due diligence from BA and ultimately (in my perspective) took their side. £20m is a meagre amount, considering the credit score impact on 420,000 people.

To some extent, the audacity of e-criminals is something we could all take as an example. If regulators keep being soft-handed, and if boards and C-levels are not made accountable for breaches that are their responsibility, then criminals will keep doing what they are doing.

Someone should be in the boardroom waving the red flag and getting everyone else to pay attention to the severity of cyber risks.

The value of a business depends largely on how well it guards its data, the strength of its cybersecurity, and its level of cyber resilience.

As I write these words, nation-states are bumping their e-horns against each other. Every nation's "worth" can also be read as both its offensive and defensive capabilities in cyberspace.

Companies will be considered cybercannon fodder and will, unfortunately, be dragged along in a huge flow of collateral damage.

Businesses need to do better. I remember someone telling me (as a justification for a lack of cybersecurity budget increase) that *"We are secure enough."*

No one is secure enough. Don't even think about it.

Cybersecurity is a never-ending chapter of a book where we are all characters with a role to play.

Acknowledging this gap is the basis for any company's Security Roadmap. Unfortunately, it has always been a struggle to make boards, C-level and execs admit that they need to do more and better.

This lack of humility is, without a doubt, the biggest challenge in Information Security today. And it will stay that way for a while.

The InfoSec Industry could unite. Sure, there is a lot of information being shared about new vulnerabilities and how to patch or fix them (when possible). But the industry is heavily segmented. I can see that even in some of the "private" InfoSec groups I am a part of. You have the quiet members who are just group voyeurs, the ones that are always typing something, even if it is banal and adds nothing to the conversation, and those who only type a message when they want to ask for something.

I have no ill feelings towards any of them, but I also have no illusions – we are all human and, as such, flawed.

Could Artificial Intelligence and Machine Learning be the best way forward?

Maybe.

But there is one single point of failure – the humans behind it. We are still fighting for Quantum Supremacy, and even if we create a machine that wants to annihilate us, *Terminator*-style, we still hold all the keys.

I believe that ML and AI will have a huge role to play, but, stepping backwards to the root of it all, the biggest risk will not cease to exist (or, at least, not soon, hopefully) – humankind.

That does not mean that we are doomed. We will continue to adapt and overcome, with the usual and expected risks of global annihilation. Maybe when we stop talking about Digital

Transformation, we will have another Digital Revolution. Oh, wait. We are still undergoing a Digital Revolution. It never stopped. So, let's keep at it.

Newer biometric authentication systems will eventually replace passwords, which is a good thing, but it won't eliminate the risk for criminals to skim irises and fingerprints. It will become a case of *"It's what you do with it that matters,"* and society does not have a good track record regarding secure implementations of . . . anything!

It is likely that companies will start storing images that will be as (if not more) important than all the other Personal Identifiable Information (PII) that we already know is sometimes left unchecked and thus easily compromised.

In every respectable company, "lessons learned" from all IT incidents is a process that is part of a "root cause analysis" process. However, most of the time, it is just something that gets written, is not shared with everyone, nor is part of a new joiner's onboarding process, so it will not serve its purpose.

Companies need to be humble enough to understand that every single breach (not just their own) can provide a valuable lesson for everyone.

Looking into the past, to some extent, is the single most valuable piece of information that companies have at their disposal.

But ask yourself, dear reader: how many companies do you know that look into other companies' breaches and aggregate information that can be used for themselves? Not many.

Companies tend to feel a lot of *Schadenfreude* (pleasure derived by someone from another person's misfortune) without realising it could happen to them tomorrow.

As I mentioned, the biggest fallacies I have experienced, the *"We are secure enough"* statement and the *"It won't happen to us, hackers do not care about us,"* come down to not being a question of caring; it's a question of opportunity.

As part of risk management, which I have been doing a lot of for the past decade, the most effective way that I have found in order to get the message across to execs and boards is actually very simple:

· · · · · · · · · · ·

No one will ever have zero risk and/or all the vulnerabilities fixed. That's unachievable.

· · · · · · · · · · ·

As such, the mitigation of said risk while adding as much complexity for a criminal while not falling down the rabbit hole of "security through obscurity" is definitely doable.

What does this mean in business terms?

You need to know where your data is. All of it. Not just PII, because that's what triggers fines and usually gets more people focused on it. Maybe non-PII data is not a harbinger of fines, but it can be a catalyst for a breach. So, keep an eye on that.

Minimising attack vectors and risk should be part of all processes a business has, in IT terms. You have to care about everything, as challenging as that is.

Then, assuming you already have products and processes in place, the last and most important hurdle is people.

InfoSec is about people as much as it is about tech. Most of the time, tech is easier to understand than people, but people are easier to manipulate.

Social engineering has taken a front-row role in hacking for one simple reason – hacking a person is easier than hacking technology nine out of ten times.

But calling every employee "the weakest link" just because they are human beings is, in my opinion, a bit too arrogant and condescending.

Companies need to understand that their weakest link could be their strongest one – if enough and proper attention is given.

Having a workforce with what I like to call *"a minimal and healthy amount of fear and respect"* regarding hackers and phishing emails is something all companies should aim to have.

And that, dear reader, is priceless.

Throwing money and products at problems has been another common mistake I have encountered. But, as much as Security is everyone's role (yes, it is), lack of training can easily create a lot of *"We are secure enough"* assumptions.

And when I say "people," I don't mean just the InfoSec team. I literally mean everyone.

Another common fallacy is that execs are too busy to watch training videos, thinking that they're only doing it for compliance.

There is only one word to define any sort of InfoSec training correctly – mandatory. And that means everyone.

The almost funny fact about execs evading training is that, by doing so, they are not only increasing the overall company risk of a breach, but they can also be tagged as negligent after an incident. It is a very strong word, but if "due diligence" doesn't work, then it's the only remaining adjective that can be applied.

Startups are another nemesis of InfoSec, and many are easy wins for hackers. The common adage of *"We're too busy to do Security,"* or *"We're too small to hire a Security team to deal with all vulnerabilities"* can only have one logical outcome. If you keep cutting corners like that, there is a very high probability that you will never grow because a breach will happen and be devastating for the business.

This is another extremely unpopular conclusion, albeit proven time and time again.

Here's another interesting statistic (yes, another one, sorry): 60% of small businesses fold within six months of a Cyber Attack.[38]

I won't discuss the veracity of the statistic itself, but even if it's 50% or 40% . . . it's a lot. That can't be disputed.

I won't go as far as stating humanity is determined to end itself, InfoSec-wise, but sometimes I am tempted to.

We mindlessly click.

We overshare.

And businesses are everything but honest when reporting their breaches and data leaks. Yahoo reported a breach three years after it happened, and then went from one billion compromised accounts to three billion.[39] What's a couple billion more, huh?

For several reasons, the nation-state and state-sponsored Advanced Persistent Threats (APTs) are not going away anytime soon.

Mainly, there are no consequences to what they do. The most devastating thing that might happen is that, well, their attacks are thwarted, but they are still criminals at large.

Secondly, well, they're persistent, right? That means that it is very likely that the full extent of the threats is still being discovered, and many (hopefully not too many) are still undetected.

Interestingly, Wikipedia only lists nine countries (China, Iran, Israel, North Korea, Russia, Turkey, United States, Uzbekistan and Vietnam) in their APT list,[40] but that's very likely to be far from the reality. Many other super-nations surely have their counter-intelligence teams, but as they are *"the good guys"* in this cyber war, they can't be listed as APTs, because they are *"fighting the good fight."* More specifically, I'm thinking UK, France, India and Germany.

Another interesting aspect about APT lists is that, depending on which site you're looking at, their names either change, or they don't even have a name or a sequential number. That just tells me that the list is not official, which makes sense, but much of it is aspirational and has a lot of "maybes."

So, without further ado:

APT1, PLA Unit 61398, Comment Panda, China
APT2, Putter Panda, China
APT3, Gothic Panda, China
APT4, Maverick Panda, China
APT5, Keyhole Panda, China
APT6, 1.php Group, China
APT7, China
APT8, China
APT9, Nightshade Panda, China
APT10, Menupass Team, China
APT12, Numbered Panda, China
APT14, Anchor Panda, China
APT15, Vixen Panda, China
APT16, SVCMONDR, China
APT17, Deputy Dog, China
APT18, Dynamite Panda, China
APT19, Deep Panda, China
APT20, Twivy, China
APT21, Zhenbao, China
APT22, Barista, China
APT23, China
APT24, PittyTiger, China
APT25, Vixen Panda, China

APT26, China
APT-C-26, Labyrinth Chollima, North Korea
APT27, Goblin Panda, China
APT28, Fancy Bear, Russia
APT29, Cozy Bear, Russia
APT30, China
APT31, Zirconium, China
APT32, Ocean Buffalo, Vietnam
APT33, Elfin, Refined Kitten, Iran
APT34, Helix Kitten, Iran
APT35, Charming Kitten, Iran
APT-C-35, Viceroy Tiger, India
APT36, Mythic Leopard, Transparent Tribe, Pakistan
APT-C-36, Blind Eagle, Colombia
APT37, Ricochet Chollima, Lazarus Group, North Korea
APT38, Stardust Chollima, Lazarus Group, North Korea
APT39, Remix Kitten, Iran
APT40, Leviathan, China
APT41, Wicked Panda, China
APT-C-43, Machete, LATAM

NB: Over the years, APT 11 and APT 13 were merged into other groups and subsequently deprecated.

Worthy mentions:

Double Dragon, China
Unit 8200, Israel
Equation Group, USA
admin@338, China
Ajax Security Team, Iran
Allanite, Russia[41, 42, 43]

You can see that the US Equation Group, described as *"one of the most sophisticated cyber attack groups in the world,"*[44, 45] does not have an APT number. Trying to fly under the radar, I see.

Well, 500 malware infections by the group in at least 42 countries[46, 47] means it's a bit too late for that.

THE APT PANDA AND APT BEAR IN THE ROOM

Criminals have a long history of conducting cyber espionage on China's behalf. Protected from prosecution by their affiliation with China's Ministry of State Security (MSS), criminals turned government hackers conduct many of China's espionage operations.

China has strength in numbers, education and training.

Their National Cybersecurity Talent and Innovation Base[48] is located in Wuhan (of all places!). It includes all of the Base's components while being capable of training and certifying 70,000 people a year in cybersecurity.

There is an expression in China: *"a child from another family,"* which represents an ideal kid who is better than you in every way. Young people will hear the "legend" of this kid from their parents, teachers, and relatives. After telling the story, they always tell you that one should get good grades like him, be talented like him, and get as many prizes as he gets.

They create peer pressure by creating a fake kid, but they don't teach HOW to be this kid.

So, all they know is competing with others, while they don't care how to win a competition.

If they can win a game without effort just by using cheats and hacks, yes, of course, they will. The majority of their young generation doesn't care about the honour or the way they win; they just care that they win.

This has become a mindset.

As such, hacking presents itself as a job opportunity, as a result of "excess production capacity." In the last decade, China experienced the explosive development of the Internet, and a Major in Computer Science was a popular option in universities.

However, as the bubble burst, many programmers were not hired by mainstream companies and became unemployed.

You can imagine how fairly easy – or, should I say, accessible – it is for them to hack something by using their knowledge.

They need to survive, so they choose to degenerate.

Let's now focus on the cold North.

Russia possesses one of the best mass-education systems in the world and has a long-standing tradition of high-quality education for all citizens. Russia's education system produces a 98% literacy rate.

But if one thinks, *"Why hackers?"* the answer is not engineering-focused education in colleges as much as math and science-focused education in schools.

Regardless of what image you have of the Soviet Union, most of them are grateful to Soviet educators for what they did, especially in the field of mathematics and physics.

The curricula in Russian schools are much more demanding than in US ones, especially in high schools.

It is not uncommon for a maths programme to include, on top of the standard stuff: limits with epsilon-delta definition and proofs, derivatives and integrals, simple differential equations, combinatorics, mathematical logic, irrational numbers, Taylor series, beginnings of complex analysis (Euler's formula), just to name a few of the things that make me cringe.

And that can be applied to other disciplines, too – like biology, physics or chemistry.

Heck, even astronomy is a standard high school class.[49]

As a result, college enrolment represents a relatively smooth transition (as opposed to the US, where a high school graduate from an average school can be in for quite a shock unless there's a sports scholarship involved), and in good universities, learning picks up at a greater pace.

It's not uncommon to study linear algebra, advanced calculus and theoretical mechanics in the first year of college.

A lot of top coders are 15 to 20-year-olds, and their school education (especially maths) plays a big role in how fast and deeply they understand and assimilate Computer Science material.

Amazingly enough, there are still those who think that all (!) technology will evolve to be secure by default and that *"Cybersecurity won't be such a major concern in two years' time,"*[50] which will potentially make a lot of cybersecurity professionals redundant, and hackers meaningless.

I believe I would only be able to fully explain and justify how these views are utterly delusional if I went the War and Peace way and wrote 1200+ pages about it.

There are two major flaws in this crystal-ball-like theory:

First and foremost, it is virtually impossible for all software to be secure. And even if it was, at one point in time, new software brings new threats and new Governance challenges.

Last but not least, cybersecurity is not just about products. We need to add two other elements into the mix: processes and people.

The combination of these three (products, processes, people) will never – ever – have zero risk.

It is obvious that *"Organisations need to embed network and cloud security from the ground up,"* but that doesn't eliminate the need for cybersecurity. On the contrary, it means that it needs to be part of the business as a whole, and that's not less work.

The fact that this "article" was written by the CEO of an IT company just highlights how easy it is to shove trash into the internet.

But back to Governance challenges.

The international environment is hardly conducive to discussions on how best to coordinate responses to the complex, cross-border dilemmas emerging around both cybersecurity and new technologies.

The international community is notoriously slow at adopting new rules and institutions to deal with new challenges, and the quandaries posed by questions of national sovereignty and democratic legitimacy are persisting. In contrast, big companies appear to be racing ahead, intent on shaping the "science, morality and laws" of new technologies such as AI, with limited public debate underpinning or guiding their efforts. In this case, "big" is obviously not "better."

Many of these same companies and the technologies they produce or exploit are increasingly viewed as instruments of state power, a fact that only adds to these sovereignty and legitimacy-related questions.

Meanwhile, growing strategic competition between the world's leading powers, especially in high-tech sectors, does not bode well for multilateral efforts to respond cooperatively and effectively. Such a competitive landscape is contributing to regulatory fragmentation and will likely delay much-needed normative and regulatory action.

The resulting trust deficit between countries poses a significant threat to international peace and security, one that existing political institutions are not necessarily prepared to handle.

Throughout history, new challenges (including those relating to technology and governance) have generally opened new opportunities and channels for cooperation. Today is no different, although the challenges at hand are highly complex and are emerging at a time of systemic political change and a rising sense of conflict and crisis. More meaningful dialogue and cooperation – however difficult – on how technological developments are affecting societies and the uses and applications of technology generating the most disruption and contestation are urgently required. Such an approach would likely afford greater legitimacy to emergent governance efforts, while also tethering them to the common good.

In national terms, states are under increasing pressure to ensure that government agencies, cybersecurity firms and researchers discover and disclose cyber vulnerabilities in a timelier fashion and prevent these vulnerabilities from being illicitly traded or otherwise misused.

While a principal aim is to strengthen transparency and oversight of government use of discovered zero-day vulnerabilities, there are concerns that such processes are bureaucratically complex and expensive and might remove pressure on companies to produce more secure products and services. Moreover, explicit processes for managing vulnerabilities could be seen as legitimising government hacking.

In light of persisting cybersecurity risks, governments also are moving toward more regulatory-focused solutions, many of which stop short of formal regulation. For instance, in 2018, the EU adopted a broad instrument called the Cybersecurity

Act. It includes a voluntary certification framework to help ensure the trustworthiness of the billions of devices connected to the Internet of Things underpinning critical infrastructures, such as energy and transportation networks, and new consumer devices like driverless cars.[51]

The framework aims to *"Incorporate security features in the early stages of their technical design and development (security by design), ensuring that such security measures are independently verified and enabling users to determine a given product's level of security assurance."*

The effectiveness of such initiatives has yet to be gauged, although sceptics (me included) often point to challenges around voluntary certification schemes in other sectors. For instance, a scandal involving the automobile manufacturer Volkswagen (an incident commonly referred to as Dieselgate)[52] showed the limitations of one such voluntary scheme. In such cases, the objectives may be good, but inherent conflicts of interest in process design, monitoring and oversight tend to undermine these goals in the longer term.

The Cybersecurity Act follows on the heels of the EU's General Data Protection Regulation (GDPR), which seeks to bolster EU citizens' data privacy and harmonise data privacy laws across Europe. The 2016 EU Directive on Security of Network and Information Systems is the first piece of legislation on cybersecurity that the EU has adopted. In the United States, there is increasing pressure on companies to prioritise consumer protection and citizen safety, plus to introduce *"proactive responsibility and accountability into the marketplace,"* including through product liability. Such an approach might be particularly useful when security flaws are easily prevented *"by commonly accepted good engineering principles."*

Even Microsoft has promoted norms of responsible behaviour for both state and industry actors, and the company has reportedly responded more positively than other corporate peers in terms of complying with new regulations such as the EU's GDPR.[53]

The firm has also raised the idea of a "Digital Geneva Convention" – a binding instrument that would protect users from malicious state activity.[54]

Along with several other industry leaders, Microsoft has also announced the Cybersecurity Tech Accord, which advocates for increased investment in (and heightened responsibility for) cybersecurity by leading industry actors.

In recent years, the announcement that has produced the most headlines is Facebook founder Mark Zuckerberg's call for greater government and regulatory action, notably in the areas of *"harmful content, election integrity, privacy, and data portability"* following the March 2019 attacks in Christchurch, New Zealand.[55]

Importantly, he stressed the need for more effective privacy and data protection in the form of a *"globally harmonised framework,"* urging, somewhat ironically, that more countries adopt rules such as the EU's GDPR as a common framework. Zuckerberg's opinion piece received a lukewarm reception, and many experts remain sceptical of his intentions.[56]

Despite this progress, significant governance challenges remain for cyberspace. Efforts to not only protect data, privacy, and human rights online but also attend to national and international security concerns are improving in some cases. For instance, according to one assessment, the EU's GDPR provides much stricter guidelines and *"strict security standards for collecting, managing, and processing personal data."*

But the instrument does provide exemptions for data controllers or processors when it comes to *"national defense, criminal investigations, and safeguarding the general public."*[57]

Progress remains much more limited or has even regressed in other countries and regions.

Despite concerns about the growing scale, economic and societal costs, and other risks of online criminal activity, states have not been able to (and likely will not) agree on a common framework for dealing with cybercrime or other malicious online activity that imperils users and hampers economic growth and development. This state of affairs is unlikely to change, given that some states continue to insist on the need for a common framework, while others remain wedded to the expansion of the existing Budapest Convention.

There are other challenges too. Progress remains slow in terms of achieving public and private sector commitments to bridge existing technological divides and move the digital transformation agenda forward. Inequalities within and between states (and cities) are growing even as technological advances continue to be made. This situation may make it even more challenging to meet the UN Sustainable Development Goals.

Some countries will further challenge modalities of internet governance, particularly states that view greater state involvement in internet governance as crucial to national security. These divergences over how the internet should be governed continue to foment tensions among states and other stakeholders. Meanwhile, several countries have announced they will seek to build their own national alternatives to the global internet, possibly further fracturing an (already fractured) world wide web. However, some observers have questioned the feasibility of such alternatives.[58]

Meanwhile, some other countries – and some non-state actors – appear to remain committed to a binding international treaty. Yet the likelihood of such a treaty is perhaps slim, not least because the key actors crucial to any agreement view cyberspace and cybersecurity in very different strategic terms. At present, there appear to be limited incentives to agree on a new regime.

How countries hold each other accountable for violating norms is just as important. Indeed, some states' persistent misuse (and potentially lethal use) of IT is driving a dangerous security dilemma involving tit-for-tat activities that have significant escalatory potential. Beyond the fact that such activities raise serious questions about the rule of law, most related crisis-management or confidence-building mechanisms would likely prove ineffective in the event of escalation if there were no real channels of diplomatic dialogue between key states. Currently, such channels are largely non-existent.

The growing number of initiatives aimed at fostering greater cybersecurity and stability do not (and perhaps cannot) deal with some of the structural issues driving insecurity and instability. This is particularly the case regarding IT products and services, which remain highly vulnerable to exploitation by actors with malicious intent.[59]

Greater and more participatory dialogue on the nature of global IT market trends and the structural levers for making IT products and services more safe and secure is urgently required. It should not be inhibited by the growing (and valuable) focus on the so-called Vulnerability Equities Processes (VEPs)[60] and other similar measures.

Finally, existing threats and vulnerabilities will surely be compounded by new problems. This means that conceptions of security will need to be reconsidered over time, and existing normative and governance frameworks will likely need to be adapted.

For instance, new threats and vulnerabilities related to the Internet of Things are emerging. As the lines between human agency and *"smart agent-like devices"* become increasingly blurred, the safety and security of related services and devices remain serious problems.

Likewise, new threats are also developing in relation to critical systems dependent on AI (such as the growing number of sectors and industries reliant on cloud computing), critical satellite systems, and information and decision-making processes, which are increasingly susceptible to manipulation for political and strategic effect. Heightened strategic competition and deteriorating trust between states further compound these challenges.

More than ever, countries need to invest in diplomacy to foster greater dialogue, cooperation and coordination on the IT-related issues that pose the greatest risks to society.

VOX POPULI

Before I shed light on how a Latin phrase meaning *"the majority of the voice of the people"* can be applied to cyberspace, I need to start by saying that even though it is most commonly pronounced as (*vox pop u leye*), it is supposed to be (*vox pop u lee*). That and "per say," which is a bastardisation of the original "per se," are two of my pet peeves.

With that out of the way, we can focus on the voice of the people.

I recently saw a post on Twitter (no, I wasn't browsing, a friend shared the link) where an InfoSec regular or "influencer" (I'm trying not to say bad things, but my initial thoughts were *"loud, maybe too loud"*) shamed an email she received from a Cybersecurity product salesperson. It tried to relate on the basis of *"Women are underrepresented and underestimated, but we're capable, so if you need me, I'm here."*

Let's dissect this for a moment.

Using – any – gender/race/religion word to try and relate with anyone as a sales pitch is wrong.

But for a known person in the InfoSec world to spend the time to put a screenshot on Twitter, share it with an 80k following, along with a verbose rant on the person, the company, and their ethics? That's also wrong.

Anyone can have a bad day and send an awkwardly wrong email by mistake, not having read it at least three times before pressing send. It happened to me. Fortunately, it was a company internal email, and the receiver didn't share it with the world. Yet I still apologised. In person.

But the example above shows how the "herd mentality" is dangerous. And I say "herd" with a love of all groups of animals, especially hoofed mammals, that don't look at Twitter. They don't know how lucky they are.

But the herd that does have Twitter are weaponising it, as per the example above.

If vox populi decide to hate on someone, it is the closest thing the internet has to stoning. In the end, it becomes social engineering and manipulation, even if unintended.

Big companies like Twitter, *which are supposedly providing a platform for free speech*, only care about the ads and the other ways they find to monetise the user data. As such, the herd is a fantastic revenue stream: it works by itself.

ARTIFICIAL INTELLIGENCE

• •

Although AI research has existed for more than five decades, interest in the topic has intensified over the past few years. This highly complex field emerged from the discipline of computer science.

The classic definition of AI dates back to 1955. John McCarthy and his fellow researchers characterised artificial intelligence as *"making a machine behave in ways that would be called intelligent if a human were so behaving."*[61]

Noting that a similar counterfactual understanding of AI underpins the earlier Turing tests, AI has been conceptualised as *"a growing resource of interactive, autonomous, and often self-learning agency (in the machine learning sense . . .), that can deal with tasks that would otherwise require human intelligence and intervention to be performed successfully."* Simply put, AI can be viewed as a *"reservoir of smart agency on tap."*[62]

AI encompasses numerous sub-disciplines that include natural language processing, machine inference, statistical machine learning and robotics.[63]

Certain sub-disciplines, such as deep machine learning and machine inference, are often seen as points along a continuum on which progressively fewer human beings are required in complex decision-making.[64]

Some observers believe this will eventually lead to artificial general intelligence or super-intelligence that either achieves or surpasses human intelligence.[65]

Yet it is fiercely debated whether AI will ever actually achieve or exceed such cognition and abstract decision-making capabilities.[66]

Nonetheless, advances in the various AI subfields are expected to bring about great economic and social benefits. Communications, healthcare, disease control, education, agriculture, transportation (autonomous vehicles), space exploration, science and entertainment are just a few of the areas already benefiting from breakthroughs in AI.

Yet the risks inherent in the ways these technologies are researched, designed and developed, and how they might be used, can just as easily undermine these benefits.

The immediate risks and challenges include the expansion of existing cybersecurity threats and vulnerabilities into increasingly critical AI-dependent systems (like cloud computing); unintended or intended consequences as AI converges with other technologies, including in the biotech and nuclear domains; algorithmic discrimination and biases; weak transparency and accountability in AI decision-making processes; overly narrow ways of conceptualising ethical problems; and limited investment in safety research and protocols.

Meanwhile, policymakers are now fixated on predictions about how automation will transform industries, the labour force, and existing forms of social and economic organisation. Predictions that automation and advanced machine learning may exacerbate economic inequalities, in particular, have stoked anxiety. Several studies on subjects like the future of work, the future of food, and even the future of humanity seek to allay these concerns while also highlighting and forecasting risk.[67]

Different AI applications and models derive (or will derive) much of their power from large quantities of collected, stored and processed online data. Concerns over data protection, privacy, and other principles and values such as equity and equality, autonomy, transparency, accountability, and due process are growing. The dual-use nature of AI applications also makes it difficult to constrain their development and regulate their use.

Moreover, recently world leaders, including Chinese President Xi Jinping, Russian President Vladimir Putin, and US President Donald Trump, have made public declarations painting AI in terms of national power projection. This trend suggests the development and use of such technologies will be complicated by growing strategic competition (in geopolitical, military, economic, and normative terms).[68]

Moreover, some countries' desire to use AI as a *"critical enabler and force multiplier for capabilities across all aspects of military power"* is becoming increasingly evident.[69]

In short, further advances in AI will likely significantly alter the contours of economics, socio-political life, geopolitical competition and conflict. According to some observers, the technology may even pose existential risks. Yet there is still time to think seriously about AI and develop stronger responses to the challenges ahead.

As of right now, with no rules and no process, humankind is risking *euthanasAI*.

IKEA? REALLY?

Back in 2010, when 3D printing was at the peak of the hype cycle, activists from the Swedish Pirate Party showed up at an IKEA trade fair. They solemnly announced that it was only a matter of time before 3D printing would disrupt the furniture industry, just like had happened to the record industry after Napster. The ability to hack furniture would soon be in the hands of the people. Therefore the multinational corporation, with its questionable right-wing political connections, exploitation of labour and environmental impact, was doomed.

Fast-forward to 2018, when IKEA started commercialising its *Delaktig* line of sofas. The new line was meant to be a modular "platform" which allowed customers to perform "furniture hacking." The new sofa was arguably inspired by the practices of *"modifications on and repurposing of" IKEA furniture that are fostered by websites such as IKEA Hackers and in dedicated furniture hacking meetings."*[70]

The example above speaks of the failures of techno-determinism as a political ideology. But the main reason it was mentioned is that it exemplifies how rhetoric, practices and innovations coming from hacker cultures can be adopted by the corporate world and repurposed towards its own goals.

The very term "hacker" seems to be losing any meaning when IKEA uses it to commercialise its furniture or when Facebook's address is *"*1 Hacker Way, Menlo Park, CA."*

The big shift in Hacker Culture has been an increased freedom of information brought on by ubiquitous internet access. Until the late '90s, you either had to figure it out on your own, know someone who knew how to hack, or be lucky enough to stumble on a 'zine or small press book. Hacking worked much like a medieval guild. Hacking knowledge was handed down and passed around. It was arcane, jargon-laden, and often wrong. The notion of hackers as "wizards" conveys the value and power of knowledge in pre-internet hacker culture, but it also represents the secrecy.

Although the cybersecurity community will likely have to live with the term and practice of hacking being associated with security issues, there has at least been a general willingness to separate the good from the bad with labels like White Hat,

Black Hat, or Grey Hat being used as a way to identify different types of hackers according to their motivations and intentions.

Hacking isn't going away, but cybersecurity professionals can once again hold their heads high when someone happens to call them a hacker.

The journey has only just begun.

CONCLUSION

· · · · · · · · · · ·

There has been an enormous transformation in the way hackers are viewed.

It is not merely a matter of a change in evaluative judgements of hackers and hacking, but in the very meaning of the terms. Hacking is now imbued with a normative meaning whose core refers to harmful and menacing acts. As a result, it is virtually impossible to speak of, let alone identify, the hackers that engage in activities of significant social value. Because the old hackers eschewed the centralisation of authority and invasive property boundaries, the shift is convenient for those who seek to establish control in the new order and economy of cyberspace.

Not only does it vilify early hackers by association with evil hackers, but it becomes virtually impossible even to perceive them, for we have lost the vocabulary with which to identify them. As a collateral loss, it is harder to deliberate over the conflicting substantive principles.

Concepts carve the world into meaningful chunks and serve particular ends, whether they are explicitly crafted or emerge naturally as the meaning of everyday language.

· · · · · · · · · · · ·

"Social reality is created by us for our purposes and seems as readily intelligible to us as those purposes themselves."

· · · · · · · · · · · ·

In the extreme, the evolution of appropriate conceptual schema may even be seen to serve the flourishing of a species. For example, in the case of monkeys, that can warn troop members about the presence of predators with special "words" conveying something about the nature of these predators – whether airborne (say, an eagle) or terrestrial (a snake).[71]

In this sense, our concepts are teleological (as in, relating to or involving the explanation of phenomena in terms of the purpose they serve rather than of the cause by which they arise), not only shaping our thoughts and utterances but facilitating, making awkward, or even depriving us of the facility to think and talk about certain things.

In some cases, such as the refined conception of predators, these conceptual schemas serve shared or common ends within a community of agents, thinkers, and speakers. But this is not universally true of all conceptual and classification schema which, as discussed above, may favour some members' interests at the expense of others.

In this way, by skewing the meaning of hackers, established institutions of cyberspace have enlisted the power of conceptual schema in their quest for order and control. The recognition of contested ends is partly what leads to the following conclusion:[72]

· · · · · · · · · · · ·

"One of this book's central arguments is that classification systems are often sites of political and social struggles, but these sites are difficult to approach. Politically and socially charged agendas are often first presented as purely technical, and they are difficult even to see."

· · · · · · · · · · · ·

We are not all well served by the transformation of "hacker" into a category which includes only at its edges those who espouse the hacker ideology (or "hacker ethic"). These hackers have much to offer to individual users of cyberspace and, ultimately, to contribute to the public good.

Nevertheless, for many of the institutions invested in strong property rights and traditional ordering, even these hackers constitute a threat. They challenge institutional strongholds and are sufficiently skilled at manipulating the underlying technologies to meet their ideological commitments. All the better if this irksome group and its causes would fade from public consciousness into the margins of a larger category typified by vandals, terrorists, and criminals. All the better for the institutions if they can craft an enemy in common with individual users and consumers so as to subordinate all who might challenge them.

Computers and the internet have extended our modes of association, action, expression and access to information, and have conjured many wondrous entities and interactions into existence. The precise nature of these entities is not always understood, and questions about them arise that have implications for policy and values.

Questions such as:

- What is a border in cyberspace?
- Where are the edges of a hypertext document?
- What is it to be an owner of something online?
- What is public; what is private?
- What is identity online; what are identities?
- Is virtual friendship, friendship, virtual war, war, and virtual sex, sex?

And what does "hacker" really mean to you, dear reader?

NOTES AND REFERENCES

· · · · · · · · · · ·

1 http://catb.org/esr/writings/hacker-history/hacker-history-3.html
2 http://www.catb.org/~esr/jargon/html/H/hacker.html
3 https://w3.pppl.gov/~hammett/work/2009/AIM-239-ocr.pdf
4 https://www.oreilly.com/library/view/the-cathedral/0596001088/ch01.html
5 Donn B. Parker, et al, Computer Crime: Computer Security Techniques (US DOJ Bureau of Justice Statistics and SRI International, 1982).
6 "The Constitution of a Hacker," 2600 Magazine (March 1984 V1 3), https://store.2600.com/products/1984?variant=1229064701
7 http://telnetbbsguide.com/
8 https://en.wikipedia.org/wiki/FidoNet
9 https://scenerules.org
10 https://www.theatlantic.com/technology/archive/2016/05/the-computer-virus-that-haunted-early-aids-researchers/481965/
11 http://www.phrack.org/archives/issues/65/15.txt
12 https://scenerules.org
13 https://emojipedia.org/
14 https://ibb.co/0QGXL2x
15 https://www.wired.com/2012/03/lulzsec-snitch/
16 https://en.wikipedia.org/wiki/The_Jester_(hacktivist)
17 https://monoskop.org/Hackers_At_Large
18 https://web.archive.org/web/20010610184511/http://www.hal2001.org/hal/03Topics/06program/index.html

19 https://web.archive.org/web/20010609205328/http://www.hal2001.org/hal/03Topics/secpriv/index.html

20 https://containerjournal.com/features/ibm-makes-case-mainframes-container-platforms/

21 https://www.ncsc.gov.uk/blog-post/rise-of-ransomware

22 https://cyber-edge.com/cdr/

23 https://rusi.org/explore-our-research/publications/emerging-insights/ransomware-a-perfect-storm

24 https://www.baesystems.com/en/cybersecurity/feature/the-nation-state-actor

25 https://thedataprivacygroup.com/blog/marriott-data-breach-traced/

26 https://krebsonsecurity.com/2022/02/russia-sanctions-may-spark-escalating-cyber-conflict/

27 https://www.itgovernance.co.uk/blog/data-breaches-and-cyber-attacks-in-2021-5-1-billion-breached-records

28 https://icdt.osu.edu/cybercanon/cyber-war-next-threat-national-security-and-what-do-about-it

29 https://www.wired.com/2015/02/nsa-acknowledges-feared-iran-learns-us-cyberattacks/

30 https://www.mandiant.com/resources/blog/fireeye_responds-wave-desctructive

31 https://www.wired.com/story/russia-false-flag-hacks/

32 LinkedIn, 18th of August post

33 https://www.wsj.com/articles/facebook-parent-meta-platforms-cuts-responsible-innovation-team-11662658423

34 https://www.propublica.org/article/machine-bias-risk-assessments-in-criminal-sentencing

35 https://www.tandfonline.com/doi/abs/10.1080/019722497129061

36 https://www.upress.umn.edu/book-division/books/technoculture

37 https://www.businessleader.co.uk/1-in-3-firms-experience-cyberattacks-weekly-new-report-finds/

38 https://www.inc.com/joe-galvin/60-percent-of-small-businesses-fold-within-6-months-of-a-cyber-attack-heres-how-to-protect-yourself.html

39 https://www.upguard.com/blog/biggest-data-breaches

40 https://en.wikipedia.org/wiki/Advanced_persistent_threat#APT_groups

41 https://www.mandiant.com/resources/insights/apt-groups
42 https://attack.mitre.org/groups/
43 https://adversary.crowdstrike.com/
44 https://web.archive.org/web/20150217023145/https://securelist.
    com/files/2015/02/Equation_group_questions_and_answers.pdf
45 https://securelist.com/a-fanny-equation-i-am-your-father-stuxnet/
    68787/
46 https://arstechnica.com/information-technology/2015/02/how-
    omnipotent-hackers-tied-to-the-nsa-hid-for-14-years-and-were-
    found-at-last/
47 https://www.pcworld.com/article/431905/equation-cyberspies-
    use-unrivaled-nsastyle-techniques-to-hit-iran-russia.html
48 https://cset.georgetown.edu/publication/chinas-national-cyber
    security-center/
49 https://vagcel.ru/en/the-adrenal-glands/o-vvedenii-astronomii-v-
    srednyuyu-shkolu-zvezdnye-voiny-kak-i-zachem.html
50 https://www.irishtimes.com/business/2022/09/12/why-cyber
    security-wont-be-such-a-major-concern-in-two-years-time/
51 https://www.digitaleurope.org/resources/digitaleuropes-position-
    paper-on-the-european-commissions-proposal-for-a-european-
    framework-for-cybersecurity-certification-scheme-for-ict-products-
    and-services/
52 https://seekingalpha.com/instablog/27130533-kirill-klip/4683136-
    volkswagen-dieselgate-fallout-dirty-truth-clean-diesel
53 https://techcrunch.com/2018/05/23/50-tech-ceos-come-to-paris-
    to-talk-about-tech-for-good/
54 https://query.prod.cms.rt.microsoft.com/cms/api/am/binary/
    RW67QH
55 https://www.washingtonpost.com/opinions/mark-zuckerberg-the-
    internet-needs-new-rules-lets-start-in-these-four-areas/2019/03/29/
    9c6f0504-521a-11e9-a3f7-78b7525a8d5f_story.html
56 https://www.theguardian.com/commentisfree/2019/apr/02/
    mark-zuckerberg-fix-the-internet
57 https://www.gdpreu.org/the-regulation/key-concepts/data-
    controllers-and-processors/
58 https://www.themoscowtimes.com/2019/02/21/russia-must-
    build-own-internet-in-case-of-foreign-disruption-a64578

59   http://www.unidir.org/files/publications/pdfs/stemming-the-exploitation-of-ict-threats-and-vulnerabilities-en-805.pdf
60   https://en.wikipedia.org/wiki/Vulnerabilities_Equities_Process
61   https://aaai.org/ojs/index.php/aimagazine/issue/view/165
62   https://hdsr.mitpress.mit.edu/pub/l0jsh9d1/release/8
63   https://rodneybrooks.com/forai-the-origins-of-artificial-intelligence/
64   https://www.iiss.org/publications/the-military-balance/the-military-balance-2018
65   Max Tegmark, Life 3.0: Being Human in the Age of Artificial Intelligence, 2018 https://www.wob.com/en-gb/books/max-tegmark/life-3-0/9780141981802#GOR009207760
66   http://www.unidir.ch/files/publications/pdfs/the-weaponization-of-increasingly-autonomous-technologies-artificial-intelligence-en-700.pdf
67   https://medium.com/lassondeschool/can-ai-help-feed-the-world-the-future-of-food-is-here-429e7c10290b
68   https://www.theverge.com/2017/9/4/16251226/russia-ai-putin-rule-the-world
69   https://www.lawfareblog.com/great-power-competition-and-ai-revolution-range-risks-military-and-strategic-stability
70   https://ikeahackers.net/about
71   https://psycnet.apa.org/record/1990-98585-000
72   https://kaios.net/posts/2015/07/15/democratizing-infrastructure/#fn:3

ACKNOWLEDGEMENTS

· · · · · · · · · · ·

Thanks to: my partner, without whom this wouldn't have been possible; my parents, without whom – I – wouldn't have been possible; PB for the help in a time of need; RG, DBM, CS and SK for the inspiration.

ABOUT THE
AUTHOR

· · · · · · · · · · ·

With a 300-baud modem in 1988 as the only means to communicate with what later became the internet, the word "hacker"
has been ever-present in the author's life.

Now with over 20 years as an InfoSec professional, Pedro
Borracha will share a journey through the origin of the
word "hacker" and how its meaning has changed since and
throughout the last century. *Hacking Through Time* includes
first-person examples, thoughts, and a focus on what can and
will (or not) happen next.

www.ingramcontent.com/pod-product-compliance
Lightning Source LLC
Chambersburg PA
CBHW051055050326
40690CB00006B/724